Nelson Advanced Science

Make the Grade

COPLAND COMMUNITY SCHOOL
FACULTY OF SCIENCE

A2
Biology
with Human Biology

BIOLOGY

**John Adds, Alan Clamp, Martin Furness-Smith,
Erica Larkcom, Ruth Miller**

 KU-729-840

Text © John Adds, Alan Clamp, Martin Furness-Smith, Erica Larkcom, Ruth Miller 2002
Original illustrations © Nelson Thornes Ltd 2002

The right of John Adds, Alan Clamp, Martin Furness-Smith, Erica Larkcom and Ruth Miller to be identified as authors of this work has been asserted by them in accordance with the Copyright, Designs and Patents Act 1988.

All rights reserved. No part of this publication may be reproduced or transmitted in any form or by any means, electronic or mechanical, including photocopy, recording or any information storage and retrieval system, without permission in writing from the publisher or under licence from the Copyright Licensing Agency Limited, of 90 Tottenham Court Road, London W1T 4LP.

Any person who commits any unauthorised act in relation to this publication may be liable to criminal prosecution and civil claims for damages.

Published in 2002 by:
Nelson Thornes Ltd
Delta Place
27 Bath Road
CHELTENHAM
GL53 7TH
United Kingdom

02 03 04 05 06 / 10 9 8 7 6 5 4 3 2 1

A catalogue record for this book is available from the British Library

ISBN 0 17 448309 0

Illustrations and page make-up by Hardlines Ltd, Charlbury, Oxford

Printed in Croatia by Zrinski

Acknowledgements
The authors wish to make the following acknowledgements:

The practice questions and mark schemes are based upon existing Edexcel Foundation questions. The questions should be seen as indicative of the knowledge required rather than of the layout of the test.

In past examination questions, where magnification has been given beside a diagram or photograph in the original question, this magnification has been adjusted for the size of the diagram or photograph as it appears in this book.

Every effort has been made to trace all the copyright holders, but where this has not been possible the publisher will be pleased to make any necessary arrangements at the first opportunity.

Contents

How to make the grade

Introduction

The aim of this book is to help you achieve a good grade in the Advanced GCE (A2) examination. It can be used with the two course books *Respiration and Coordination* (Unit 4) and *Genetics, Evolution and Biodiversity* (Units 5B and 5H). In the course guide, *Tools, Techniques and Assessment in Biology*, you will find more information and help with the practical work linked to the specification.

The key to success in any examination is a thorough knowledge of the subject together with the confidence to be able to select the relevant facts when answering the questions. Effective learning and systematic revision help to consolidate your knowledge. As it is possible to take the unit written tests at different stages in your course, it is important to develop good study skills from the beginning, so that you can plan ahead and not leave all your revision until the last minute.

With the new specifications, where synoptic assessment is built into the written tests for Units 5 and 6, you will need to be able to recall topics studied for Advanced Subsidiary (AS) Units 1, 2 and 3. If you are not familiar with or are unsure of the content of Units 1, 2 and 3, then revisiting them will make learning the A2 units easier.

In this book, we suggest ways of organising your work so that you can make the best use of your time. Our main aims are to:

● give you some ideas on how to learn topics

● help you to devise different ways and strategies for revision

● give you some questions to try

● offer advice and give guidance on how to answer different types of questions.

Units 4 and 6 are common to both the Biology and the Biology (Human) specifications. If you are studying Biology (Human) note that you should refer to Unit 5H (Genetics, Human Evolution and Biodiversity) rather than to Unit 5B.

For each unit in this book you will find:

● a **general introduction to the unit**, outlining its content

● a list of the **major topics**: these correspond to the subsections of the unit as set out in the specification.

Within each unit, each major topic has:

● its own **introduction**, outlining the subject content and telling you what you need to know from other units to help your understanding

● a checklist of **things to know and understand**

- a checklist of the **required practical work**
- **helpful hints** on the practical work
- **questions** and **activities** to test your knowledge and understanding
- **practice questions** similar to those that you might get in unit written tests.

The section that covers Unit 6 contains specific information on synoptic assessment, together with the coursework assessment (T2) and its written alternative (W2).

At the end of each unit, there are some **assessment questions** for you to try as you make your final preparations for the written tests. We have tried to include a wide range of different types of question style, so that you can become familiar with the different ways in which you may be required to present your knowledge or analyse data. The answers to all the questions are provided at the end of the book. In the margins and in the Answers sections, you will find *Helpful hints*, which provide advice, guidance and warnings related to your learning and revision, and *Examiners' comments*, which give explanations for some of the answers given or point out common mistakes.

Note: For Units 1–3 of the Advanced GCE in Biology see *Make the Grade in AS Biology*.

How to make the grade

How is A2 different from AS?

Many fundamental biological concepts are introduced in the AS specification and expanded in the A2. For example, before you can appreciate the details of complex metabolic pathways, such as respiration and photosynthesis, and the role of organs such as the kidney, it is necessary to have a sound understanding of the nature of biological molecules, the roles and functioning of enzymes, and the properties of membranes. Not only does the A2 specification contain topics that are more demanding, but the written tests contain fewer straightforward questions testing knowledge. For A2, as well as studying more complex topics, you need to develop more advanced skills. More of the questions will test your ability to analyse data and apply your biological knowledge to unfamiliar situations. Built in to the written tests is the synoptic assessment, where you are required to demonstrate your knowledge of the whole specification, drawing information from more than one unit.

As you work your way through the A2 units, you will notice that we have indicated, where relevant, the topics from other units that form the background or are related to the new topics you are learning. It is sensible to review these earlier topics so that you can learn the new topics more effectively.

Here are some of the major topics and concepts, which appear as foundation in the AS and are then developed in the A2.

- molecules (Unit 1) and metabolic pathways (Units 4 and 5B)
- DNA, chromosomes and the genetic code (Unit 1) and genes and heredity (Unit 5)
- adaptations of organisms (Unit 2) and interrelationships within ecosystems (Unit 5).

What topics will I study at A2?

The A2 specification consists of three units and their content is outlined in Table 1.

Table 1 *Summary of the Edexcel specification content for A2 Biology and Biology (Human)*

Unit 4 Respiration and coordination and Options
Core material – includes cellular respiration, regulation of the internal environment, response to changes in the external environment, nervous and chemical coordination.
Options – a choice of one of the following:
Option A Microbiology and biotechnology
Option B Food science
Option C Human health and fitness
Note: This unit is common to both Biology and Biology (Human)

Unit 5B Genetics, evolution and biodiversity	Unit 5H Genetics, human evolution and biodiversity
This unit includes autotrophic nutrition, biodiversity and classification of organisms, patterns of inheritance, continuity of species and speciation, succession and stability in ecosystems and applications of gene technology.	This unit includes biodiversity, succession and stability in ecosystems, patterns of inheritance, continuity of species and speciation, human evolution, human populations and applications of gene technology.

Unit 6 Synoptic and practical assessment
Paper 01: (T2) the individual study which will be partly teacher-assessed and the written report marked by Edexcel.
Paper 02: (W2) a written test as an alternative to the individual study.
Paper 03: Synoptic written paper.
Note: Candidates offer *either* Paper 01 + Paper 03 *or* Paper 02 + Paper 03
Note: 5B refers to Biology and 5H to Biology (Human)

Table 2 *Summary of the unit test papers for A2 Biology and A2 Biology (Human)*

Unit test	Mark allocation	Time allowed for written test	Types of question
4	70 (core: 40; option: 30)	1 hr 30 mins	about 9 compulsory, structured questions, worth from 4 to 12 marks, including one free-prose question
5B or 5H	70	1 hr 30 mins	about 8 compulsory, structured questions, worth from 4 to 12 marks, including one free-prose question
6 Paper 03	38	1 hr 10 mins	2 compulsory, longer, structured questions and an essay
Note: 5B refers to Biology and 5H to Biology (Human)			

Are the written tests different from those at AS?

There are three written assessment tests in the A2. In some ways they are similar to the tests in the AS, in that most of the questions are structured and your answers are written in the question-answer booklets, but there are some important differences. Table 2 summarises the papers that have to be taken, with an indication of the mark allocation, time allowed and question types.

Everyone must take the written tests. In addition, in Unit 6, you must take either the individual study (Paper 01) *or* the written alternative (Paper 02). Both these have a maximum of 32 marks. Paper 02 lasts 1 hr 20 minutes and contains 2 compulsory questions.

The questions in Unit tests 4 and 5 will be similar in style to those you have already met at AS. The shorter questions are designed to test your knowledge and understanding of the content of the specification of each unit, whereas the longer questions may also require you to show that you can interpret data presented to you in the form of graphs or tables. At least one of the questions will require you to write an answer in continuous prose. Some of the shorter questions may require a greater depth of understanding than was required for AS. For example, at AS you might have been asked to name the structures on a diagram; at A2 you might be asked to give their function. At AS, you could be asked to define a term ('What do you understand by the term facilitated diffusion?') or describe a feature, whereas at A2 you might be asked to distinguish between two terms ('Distinguish between osmosis and active transport.').

The written test for **Unit 4** has two sections: the first section contains questions relating to the core and the second section contains the questions relating to your chosen option. There is more information on the format of this paper in the chapter on Unit 4 on page 1.

There are separate question–answer booklets for **Unit 5B** and **Unit 5H**. Each booklet contains two sections: the first section contains questions relating to the content of the unit and the second section contains synoptic questions. The types of questions in the first sections are structured and similar in style to those in Unit 4. These test the content of the units. You should check carefully that you have the correct booklet for the unit you have been taught [Unit 5B for Biology, Unit 5H for Biology (Human)]. It is unlikely that you will have the depth of knowledge required to answer successfully the questions for the other unit. The Synoptic questions are designed to give you the opportunity to make connections between at least two units of the whole specification, and to use the skills and ideas you have acquired throughout the course to interpret and analyse unfamiliar data. These synoptic questions will not include any topics from the options. As with all the other written tests, at least one question will be a 'free-prose' question.

Unit 6 differs from all the other assessment units in that it does not have any specification content. The components are described in the chapter on pages 95–8, where there is more explanation of synoptic assessment and advice on essay writing. In addition, this chapter explains the requirements of the practical assessment (T2) and the written alternative test (W2).

Get to know your specification

You can download a copy of the specification from Edexcel's website. *Visit http:/www.edexcel.org.uk*

When planning how much time you need to spend on each of your subjects, it is necessary that you know how much work is involved and this can be estimated by reading the specification. The new Edexcel specification has been designed to indicate exactly what you need to know and also to give you some idea of the depth and detail required. There should be copies of the specification in your school or college, or you can get your own from Edexcel.

In the specification of each unit, you will notice that certain words occur frequently in describing the content

For any topic, knowledge which you are expected to have, either from Key Stage 4 or from the study of other units, is indicated by the term ***recall***. For example, in Unit 4, Topic 4.1 ('Metabolic pathways') 'Recall the structure of a liver mitochondrion' relates to Unit 1, Topic 1.3 ('Cellular organisation')

'Describe the structure and understand the roles of ... mitochondria ... .' You will find many instances of the use of this term within the specification and it could be helpful to keep a note of such links as a useful aid to revision for the synoptic assessment.

The term **understand** is the most frequently used and it means that you should have gained sufficient knowledge of the material to be able to explain the biological principles involved and apply that knowledge to different situations. For example, if you have an understanding of the significance of ATP in metabolism as the immediate supply of energy for biological processes (Unit 4, Topic 4.1), you will be able to explain the significance of the production and use of ATP during the stages of photosynthesis (Unit 5B, Topic 5B.1).

Know and **describe** are terms that indicate you should state the facts and give straightforward accounts of structures and processes respectively. *Explain* requires you to give reasons for events or processes. 'Explain the bioaccumulation of non-biodegradable toxins' (Unit 5B, Topic 5B.3; and Unit 5H, Topic 5H.4) would need a knowledge of the details of how and why toxins accumulate in organisms and how they are passed from one trophic level to another, with some attention to the consequences.

Terms such as **discuss** and **distinguish** indicate that you need enough knowledge of the topics to be able to give a balanced account or to recognise comparable differences between structures or processes. A good example of the use of the term *discuss* appears in Units 5B and 5H in the topic on 'Gene technology': 'Discuss the social, ethical and economical implications of the development of genetically modified organisms'. Whatever your own views and opinions, you would be expected to put forward a balanced view of the topics.

Occasionally, it is relevant to show an awareness of the significance of a process or factor without necessarily having a detailed knowledge of the underlying principles. This is indicated in the specification by the term **appreciate**. In Unit 5B, Topic 5B.1, 'Appreciate uptake by roots of mineral ions', means that you need to be aware of the general principles of uptake but not the details, which may vary depending on the ions involved. In Unit 4, Topic 4.2, 'Appreciate the differences between nervous and hormonal coordination' requires you to be able to compare the two processes in outline: you are not expected to know all the details of the ways in which different hormones act.

How can I improve my study skills?

It is very easy to get into bad habits, to put off learning class notes and leave written assignments until the last moment, but this often leads to panic just before the examination. Last-minute cramming can sometimes work, but then you probably just forget what you have 'learnt' after you have taken the examination. This has unfortunate consequences later when it comes to synoptic assessment. If you have not learnt the AS topics thoroughly, then you have to go through many of them again as they will be needed for some of the topics in the A2, and you will certainly need to be sure of the contents of Units 1, 2B or 2H, and 3 before you take the Unit 6 written test.

Organise your notes

It is important to take good notes during classes. You need to concentrate so that you do not miss vital points and links. You should then check to see that you have not omitted anything. You can fill in gaps, from your textbook or by asking your teacher for clarification. Do not be afraid to say if you do not understand. Most teachers are happy to go over difficult concepts as they are being taught so that their students gain a clear understanding of the principles from the beginning. You may not be the only student who finds a topic difficult, so you could be doing everyone else a favour by asking for more explanation. The teacher may be able to present the information in a slightly different way which could be helpful. Remember that the associated practical work can also be helpful in your understanding of some of the topics.

There are a number of different ways in which you can organise notes that you take in class and notes that you make from textbooks. It seems sensible to use headings, sub-headings, bullet points, different colours for underlining and highlighting, and abbreviations. You have probably devised your own favourite strategies, but some ideas that you might like to try out are given later in this chapter. Perhaps the most important aspect of organising your notes is the ability to summarise them, so that you can learn them effectively. Hints for such summaries are given in the *Activities toolkit* (see pages xi–xiii).

Organise your time effectively

Good study skills can become a habit. It usually helps if you break down the tasks you need to do into 'Daily tasks', 'Weekly tasks' and 'End of topic tasks'. If you can manage to do this, then revision will not be a major cramming exercise and you could avoid that feeling of panic as you realise how quickly the day of the written test approaches. A suggested outline of these tasks is given below.

Daily tasks
After each lesson, spend about 10 to 15 minutes looking through your notes to check that they are complete. If there is something that you do not understand:

● read the relevant chapter in your textbook and, if necessary, amend your notes so that the topic is clear
● discuss the problem with another student in your group
● ask your teacher at the beginning of the next lesson.

It is essential to make sure that you have a good grasp of new topics from the beginning as it is much more difficult to catch up later. In addition, many topics taught early in the course are needed as a foundation for later topics, so learning the earlier work will stop you floundering with the current topic.

Weekly tasks
It is a good idea to develop the habit of reading through your notes on a weekly basis. You may find it helpful to:
● highlight headings and key words
● add extra notes from your textbook if you think they could be helpful or explain things in a different way

- add page references for the relevant sections in the textbook so that you can quickly turn to the right page
- check that you have written up any practical investigation that you have done during the week
- complete all outstanding homework assignments.

End of topic tasks
At the end of each topic, you need to check that you have:

- a complete set of notes
- covered the whole of the topic according to the specification
- carried out and written up all the required practical investigations.

You are now in a position to summarise your notes in preparation for the learning and revision process. There are a number of different strategies that you could use in making your summaries and we give you some tips in the *Activities toolkit* on pages xi–xiii.

How can I make my revision more effective?

Every time you read through your notes or summaries, you are helping to commit them to memory, but remember that it is not sufficient just to be able to recall large chunks of notes – you also need to be able to use your knowledge and apply it to different situations. The best way of making sure that you are well-prepared is to combine solid learning with as much practice in answering different types of questions as possible. We have suggested a number of activities. These should help you to become more familiar with the information in each topic, followed by the opportunity to test yourself and a selection of practice questions.

You will need to make yourself a realistic revision timetable, taking into account:

- the date of the written test
- written tests in other subjects
- any lessons and current classwork you may still be doing (this applies particularly if you are taking tests in the January examination period)
- how long you can concentrate
- how much material you have to cover.

By now you will know how long you can concentrate without needing a break. For most people, an hour is about the maximum. Then take a break of about 10 to 15 minutes, and perhaps have some refreshment or change of the topic. Once you have made a set of revision cards, you can make use of quite short periods of time to review these, reinforcing your knowledge of the topics summarised. Where you do your revision could make a difference. Some people prefer to have no distractions while others feel the need for some background music. It is certainly a good idea to organise yourself at a desk or table, with adequate lighting and a comfortable chair.

By suggesting the activities for each topic, we encourage you to make your revision active rather than passive. If, as recommended, you have read through your notes after each lesson and every week, you should be familiar with the subject content. Then you have reached the point where you need to try out different ways of testing yourself with questions similar to those you may have in the unit (assessment).

Activities toolkit

Active revision should be more than just reading through the notes and testing yourself. You need to become actively involved, working at it constructively and building up your own summaries of different topics. You may discover fresh ways of looking at topics and making cross-connections. This approach may help you to see the underlying patterns or highlight similarities and differences and lead you to a better understanding of the subject.

There are different ways of presenting and summarising information and you need to find out which ones suit you. In the 'Testing your knowledge and understanding' sections in this book, your answers to the shorter questions will show you how familiar you are with the main technical terms and concepts. The other questions in these sections contain ideas for activities that you can use to develop your own revision cards and materials. The activities help to improve your revision techniques and your learning. By using them you can find out which ones work best for you.

To help you with this here is a range of activities that you can try:

● spider diagrams
● concept or mind maps
● annotated diagrams
● summary tables
● flow diagrams
● lists of vocabulary and definitions
● revision cards.

Spider diagram

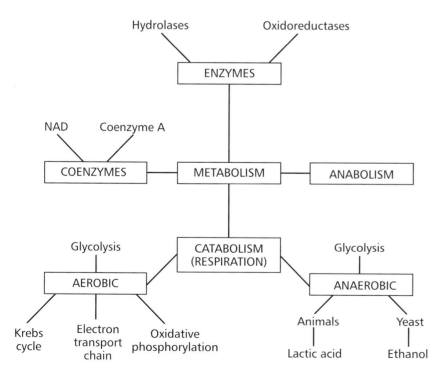

Spider diagrams and concept maps

Making up spider diagrams and concept or mind maps gives an overview of a topic or unit. The topics in each unit are a good starting point for making these. They should be structured in a clear and logical way. The diagram shown on the left is an example from Unit 4.

Annotated diagrams

Drawing diagrams of apparatus, objects or organisms should conform to a pattern that suits you. You can label them by starting at the top right-hand side and then draw lines to the parts on the outside first and the centre last. It is a good idea to add numbers or letters to the lines and to put a key below the diagram. Then you can cover up the key or diagram and use it to test yourself. If the labels, along with short descriptions, are put on the diagram, then it will become an annotated diagram.

Summary tables

You can construct tables that highlight similarities or differences to help you make connections and summarise information. The activities in the units give you several examples of suitable features to select for a table. With practice you will be able to make the selection for yourself. The table below is an example from Unit 4.

Hormone	Insulin	Glucagon	Adrenaline
Stimulus	Increase in blood glucose level (BGL)	Decrease in blood glucose level (BGL)	Sympathetic nerve impulse
Site of production	Beta cells (Islets of Langerhans in the pancreas)	Alpha cells (Islets of Langerhans in the pancreas)	Adrenal medulla (In the adrenal glands)
Site of action	Liver cells	Liver cells	Liver and muscle cells
Effect at site of action	Increased uptake of glucose by liver cells. Promotes increased conversion of glucose to glycogen (Glycogenesis)	Promotes increased conversion of glycogen to glucose-phosphate (Glycogenolysis)	Promotes increased conversion of glycogen to glucose-phosphate (Glycogenolysis)
Effect on blood glucose level	Decrease	Increase	Increase

Flow diagram

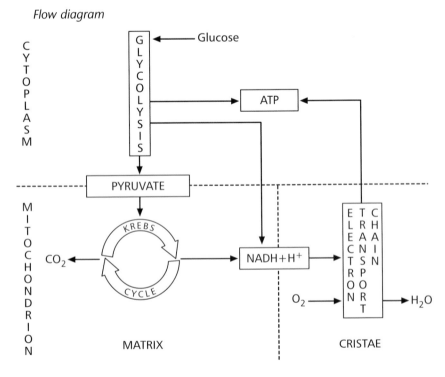

Flow diagrams

Another idea is to draw flow diagrams to show the major steps and processes involved in a procedure. This helps your understanding because you may have to simplify your notes and present the information in a different way. The flow diagram shown on the left is an example from Unit 4.

Revision cards

Some of the *Helpful hints* suggest making revision cards. Revision cards are easy to carry around and use when you have a little spare time, for example when waiting for that bus to arrive. A systematic approach to labelling these makes it easier to find the right card when you want to revise a particular unit or topic. Try to

Revision card

UNIT 5 (Genetics, evolution & biodiversity)
B.4 or H.1 Genetics and evolution
              Sources of inherited variation

Meiosis	• reduction division • 1st and 2nd division • random separations of homologous chromosomes • produces haploid cells
Chiasmata	• occur after pairing of homologous chromosomes • points of contact between chromatids • move along the bivalent (terminalisation)
Cross-over	• exchange of sections of DNA between chromatids • new combinations of alleles produced if exchange between non-sister chromatids.
Recombinants	• result of cross-over between non-sister chromatids • chromosomes have new combinations of alleles
Point mutations	• changes in base sequence due to deletion, insertion or substitution of a base • can cause frameshift
Mutagen	• chemical (eg mustard gas) or radiation (ultraviolet light) • increases rate of mutation • can cause point or chromosomal mutation.

follow the same system that you have for your notes – put the unit title and the topic at the top and then number each card for the topic. The card could have any of the outcomes of a revision activity on it. You can also use the card as a testing exercise if you use part of the card as a question and cover up the answers on the card. The example on the left illustrates this. This one has a list of vocabulary on the left-hand side and the descriptions on the right. The *Helpful Hints* will usually suggest additions that could be made: in this case the types of chromosome mutation could be added.

What questions will I have to answer on a Unit test?

On the unit written tests, apart from Unit 6, all the questions are compulsory, so you have no choices to make. Descriptions of the written tests, their mark allocation, time allowed, and number and types of questions are given in Table 2 on page vi.

The majority of the questions are of the type known as *structured*, where you are asked for specific information in each part of the question. The shorter ones (worth from 4 to about 7 or 8 marks) mainly test your knowledge of the specification content, whereas the longer ones (up to 12 marks), may ask you for more detailed analysis of data as well. Longer structured questions often begin by asking something straightforward, such as a definition or naming structures on a diagram, before presenting you with experimental data for your consideration. There is often a clear progression in difficulty in such questions: you could be asked to describe a trend in the data, then to explain it, and finally to make some suggestions about a situation, linked to the data or the experiment, that you might not have encountered previously. It is wise to look carefully at the mark allocation for each part of the question to assess the amount of information that you need to include (see page xv).

The examiners use different styles of questions. These include filling in gaps in prose passages, naming structures on diagrams, drawing or completing diagrams, completing tables by writing or ticking in the boxes, and describing processes and procedures. In all cases, the instructions given should make it absolutely clear what you are supposed to do. It is worthwhile checking, though, just to make sure. In one recent examination, candidates were asked to complete a table by writing in a one-word answer, but about half the candidates used descriptive phrases, using several words – a clear case of not reading the instructions. We have used examples of all the different question styles in both the practice and assessment questions for each unit.

The examiners use various words and phrases to guide you into giving particular answers. Some of these words are terms used also in the specification and you should have no difficulty in understanding what is meant by *explain*, *describe*, *discuss* and *distinguish between* when they occur at the beginning of a question. Other clear instructions are given by *calculate* (where some mathematical treatment of data is required), *define* (where you need to make a precise statement about a biological term), and *state* or *name* (where only a simple factual statement is needed). If you are asked to **compare**, then you should remember to include both similarities and differences in your answer. This is in contrast to *distinguish between* where you need only refer to the differences, but you must make sure that those differences refer to the same feature or process; for example, phloem contains companion cells but xylem does not; aerobic respiration yields more ATP than anaerobic respiration.

Sometimes direct questions are asked. '**Why** is the apparatus kept in a water bath?' '**Where** do the reactions of the Krebs cycle take place?' '**How** would you modify this experiment to investigate the effects of different wavelengths of light on the rate of photosynthesis?' The answers should be concise, ranging from one word to a few simple sentences stating the facts.

Perhaps one of the most difficult instructions to interpret is *suggest*, as in 'Suggest reasons for ...' or 'Suggest an explanation ...'. It indicates that you are not expected to have learnt the answer in the course of your studies, but that you should use your biological knowledge and understanding and apply it to an unfamiliar situation. One of the best ways of seeing how this works is to check the mark schemes of past questions. You can then see what sort of answer is acceptable. With questions of this style, there may not be one 'right' answer, so it is worthwhile writing down an answer, provided that you think it is relevant and it is biologically correct. It is better to write something, rather than leaving a blank space.

In Unit 6 there is an essay, but in the other unit tests you will be required to answer at least one question using continuous prose. Usually, such questions ask you to 'Describe ...' or to 'Give an account of ...', and the topic will be a fairly broad one. Similar questions in the past have been set on topics such as the events of mitosis, the structure and functions of carbohydrates, or the cardiac cycle. You are being assessed on your knowledge of the facts, so you should attempt to make your points in the correct sequence, avoid repetition and try to cover the whole range of the topic. Many candidates find it useful to jot down a very brief plan, but it is not necessary to make a very detailed one as you will be using up valuable time. Sometimes it is helpful to include a well-annotated diagram in your answer or even to use numbered points.

Last-minute preparations... and doing the Unit test

Hopefully, by the time the day of the written test arrives, you will feel confident and ready to do your best. Check exactly when you are expected in the test room. Don't turn up in the afternoon for a test that took place in the morning. It is also better to be early, rather than rushing in at the last minute. Make sure that you have all the right pens, pencils, a ruler, an eraser and your calculator, with a spare new battery.

When you get your question paper, check that it is the correct one for the unit test that you are taking. Don't waste valuable time by getting part way through the paper before you realise you are doing the wrong test.

Once you have started, it is wise to work steadily through the paper, answering each section in turn. If you come to a difficult part or where you are not sure of the answers, then make a note to go back and check this later. It is better to think *before* writing, so that you avoid too much crossing out – this can sometimes be confusing to the examiner.

- Don't take a red pen with you as the awarding body doesn't allow you to use this colour – the examiners use red for marking the papers.
- Don't use correcting fluid – it takes time to dry and you may forget to fill in the gap.

Look carefully at the mark allocation. It tells you how many marks there are for that part of the question, and also indicates approximately how much you need to write. The booklets are designed to allow sufficient space and lines for average-sized writing for each answer. If it seems you need to write more, then you are probably not keeping your answer relevant, so you should try to be more concise.

If you do need to use extra (supplementary) sheets, make sure that you write your name on them and tie them in securely at the end of the test. It is a good idea to make a note on the page of the booklet to say that the answer is continued on a supplementary sheet, so that it is not missed.

Make sure you have a little spare time at the end. Use this to make a final check of your answers, but beware of changing too much at the last minute. Your first thoughts, when you are working steadily through, are very often the correct ones, rather than a panicky afterthought as the time is running out.

4 Respiration and coordination and Options

Introduction

The content of this unit is common to both the Biology and the Biology (Human) specifications.

There are two parts to this unit of the specification: a core section on respiration and coordination, and the options. There are three options:

A Microbiology and biotechnology

B Food science

C Human health and fitness

Everyone needs to study the core section but there is a choice with respect to the options. You may be able to choose which option to study yourself, or your school or college may decide for you. Whatever the choice, when the entry for the written test is made, you will be entered for the option which has been taught in your centre.

There will be three separate question-answer booklets. Each booklet will contain the questions for the core topics and the questions for Option A *or* Option B *or* Option C, so you should check carefully at the beginning of the test to see that you have the correct booklet.

Core section: Respiration and coordination

In this unit, there is an expansion and extension of some of the AS topics, particularly those in Unit 1 and Units 2B or 2H. An understanding of metabolic pathways in general, and respiration in particular, requires a fundamental knowledge of biological molecules and the structure and functioning of enzymes. Knowledge of cell structures and the passage of molecules and ions across membranes will also help you understand how energy is supplied for biological processes. Having considered how living organisms exchange materials with their external environment in Unit 2, we now consider how the internal environment is regulated and how changes in the external environment are detected and responded to.

The core section of this unit is divided into two topics:

1 Metabolic pathways

2 Regulation of the internal environment

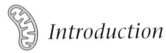

Topic ❶ Metabolic pathways

Introduction

The total of the chemical activities in a cell is called **metabolism**. It is important to have a sound understanding of the term **metabolic pathway** and to appreciate the significance of ATP in supplying energy for metabolic processes. Before studying the topic in depth, it would be helpful to review your knowledge of carbohydrates in order to understand the nature of respiratory substrates. In addition, the role of **enzymes** is important. In Unit 1, you will have studied the structure and the essential roles of enzymes in living organisms, catalysing the many hundreds of individual reactions which occur in cells. In this unit, the two key properties of enzymes which are of importance are their catalytic power and their specificity. Each small step of a metabolic pathway is catalysed by a specific enzyme: those of particular relevance in **cellular respiration** are the hydrolases and the oxidoreductases. You need to learn some details of the steps of glycolysis, aerobic and anaerobic respiration. The level of detail required is clearly stated in the specification and it is not necessary to learn all the intermediate stages of any of the processes, although most textbooks give them. You need to know the Krebs cycle in outline and to understand how ATP is generated during oxidative phosphorylation. Do not forget the structure of mitochondria, studied in Unit 1, because you need to understand where the stages of aerobic respiration take place.

The practical work associated with this topic involves experiments which illustrate the role of hydrogen acceptors. Such experiments reinforce your general knowledge of enzyme reactions and help you to understand the steps of the metabolic pathway involved. You could be asked questions on these experiments, such as describing or commenting on apparatus and interpreting experimental data. If you are unable to do the experimental work yourself, make sure that you know how the apparatus works and the nature of the reagents involved, particularly where colour changes occur.

Checklist of things to know and understand

Before attempting to answer any of the questions, check that you know and understand the following:

❑ the concept of a metabolic pathway as a sequence of enzyme-controlled reactions

❑ the roles of enzymes in the control of metabolic pathways, illustrated by oxidoreductases and hydrolases

❑ the meanings of the terms *anabolism* and *catabolism*

❑ the significance of ATP in metabolism as the immediate supply of energy for biological processes

❑ that monosaccharides are converted to pyruvate during glycolysis

❑ that during glycolysis hexoses are first phosphorylated, then broken down to glycerate 3-phosphate (GP)

❑ that glycolysis produces reduced coenzyme (NADH + H$^+$) and ATP

❏ that during aerobic respiration the events of the Krebs cycle result in the production of carbon dioxide, more reduced coenzyme and ATP

❏ the role of the electron transport chain in generating ATP by oxidative phosphorylation

❏ the role of molecular oxygen as a hydrogen acceptor forming water

❏ the structure of a liver mitochondrion, including the inner and outer membranes and the inter-membranal space

❏ the role of mitochondria as the site of the Krebs cycle and electron transport chain

❏ the location of enzymes and electron carriers, and the role of oxidoreductases

❏ that under anaerobic conditions pyruvate forms lactic acid in muscle, or ethanol in yeast

❏ the differences in the yields of ATP from the complete oxidation of glucose and from fermentation of glucose to lactic acid or ethanol.

Practicals

You are expected to have carried out practical work to investigate:

❏ *the role of hydrogen acceptors using a redox indicator, such as methylene blue or tetrazolium chloride.*

 Practical work – Helpful hints

Redox indicators are substances which are used as artificial hydrogen acceptors, because they undergo a colour change when they are reduced. One example of a redox indicator is tetrazolium chloride (TTC) which is colourless in the oxidised state, but forms a red compound when it is reduced. Methylene blue changes from blue to colourless when it is reduced.

● This investigation relies on the activity of dehydrogenases, enzymes which oxidise a substrate by removing hydrogen. The hydrogen then reduces the redox indicator, which causes the change in colour.

● An actively respiring suspension of yeast cells is mixed with a solution of TTC and incubated in a water bath.

● The time taken for the suspension to turn a standard pink colour is noted.

● This can be repeated at a range of temperatures to investigate the effect of temperature on the activity of dehydrogenases.

● If methylene blue is used as the redox indicator, it is important not to shake the tube, because the reduced methylene blue is readily re-oxidised by atmospheric oxygen.

 Testing your knowledge and understanding

The answers to the numbered questions are on page 105.

To test your knowledge and understanding of metabolic pathways, try answering the following questions.

1.1 Explain what is meant by the term *metabolic pathway*.

1.2 Describe the functions of oxidoreductases and hydrolases.

1.3 Distinguish between the terms *anabolism* and *catabolism*, giving an example of each.

Helpful hints

If you are asked to state uses of ATP, it is better to give a specific use, rather than a general reference to 'providing energy'.

The Krebs cycle is also referred to as the TCA (or tricarboxylic acid) cycle.

You are not expected to know every step in glycolysis!

Be specific here and state 'ethanol' rather than just 'alcohol'.

Mark allocations are given for each part of the questions and the answers are given on pages 105–6.

1.4 State three uses of ATP in cells.
1.5 Name the three main stages of aerobic respiration.
1.6 Draw a simple flow chart to outline the process of glycolysis.
1.7 State the products of glycolysis.
1.8 State where, in a cell, glycolysis occurs.
1.9 Draw a flow chart to show the outline of the Krebs cycle.
1.10 List the products of the Krebs cycle.
1.11 State where, in a eukaryotic cell, the Krebs cycle occurs.
1.12 Outline the process of oxidative phosphorylation.
1.13 What happens, finally, to the protons and electrons?
1.14 State the products of anaerobic respiration in yeast cells.
1.15 State the products of anaerobic respiration in skeletal muscle.
1.16 What is the net yield of ATP in anaerobic respiration?
1.17 Draw a labelled diagram of a mitochondrion and show where the Krebs cycle **and** oxidative phosphorylation take place.

 Practice questions

1   The diagram below shows an outline of glycolysis and the Krebs cycle.

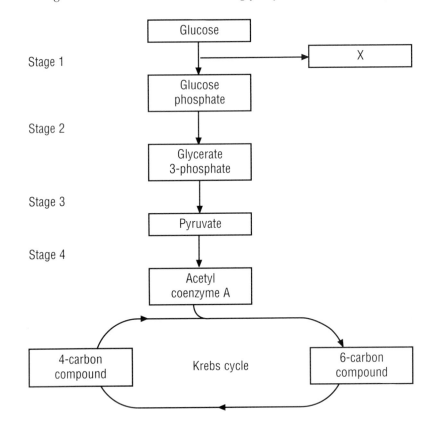

This is an example of interpretation of a flow chart, showing part of a metabolic pathway. Always read the question carefully before you start to answer it, and note the allocation of marks for each section. In part (d), there are three marks, so you should try to think of three relevant points to score full marks.

(a) Name substance X, used in stage 1.   **[1]**
(b) State where stage 4 occurs in a eukaryotic cell.   **[1]**
(c) Name **two** products of the Krebs cycle.   **[2]**
(d) Describe what happens to pyruvate under anaerobic conditions in a yeast cell.   **[3]**

(Total 7 marks)
(Edexcel 6041, B/HB1, January 1999, Q.4)

2 The diagram below shows some of the stages of anaerobic respiration in a yeast cell.

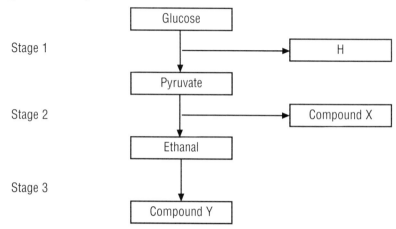

Helpful hints

Notice that this question is about *anaerobic* respiration, which affects the answer to part (b).

(a) (i) Identify compound X, produced at stage 2. **[1]**

(ii) Identify compound Y, produced by stage 3. **[1]**

(b) State what happens to the hydrogen atoms produced by stage 1. **[2]**

(c) Name **two** products of anaerobic respiration in muscle. **[2]**

(Total 6 marks)

(New question)

3 The diagram below shows an outline of anaerobic respiration in muscle.

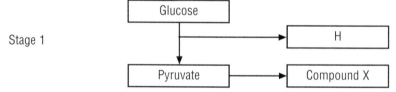

Helpful hints

Another question on *anaerobic* respiration, this time in muscle. Again, think carefully before answering part (a)(i).

(a) (i) State what happens to the hydrogen, removed during stage 1. **[2]**

(ii) Identify compound X. **[1]**

(b) Explain why it is necessary to convert pyruvate to compound X. **[2]**

(Total 5 marks)

(Edexcel 6041, B/HB1, January 1998, Q.5)

4 The diagram below summarises the electron transport chain, which is a stage of cell respiration. A, B and C represent the electron carriers.

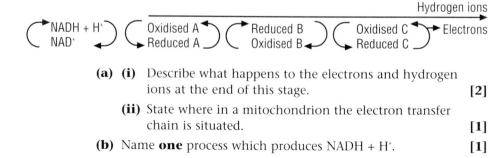

(a) (i) Describe what happens to the electrons and hydrogen ions at the end of this stage. **[2]**

(ii) State where in a mitochondrion the electron transfer chain is situated. **[1]**

(b) Name **one** process which produces NADH + H⁺. **[1]**

(c) ATP is synthesised as a result of this stage of respiration. State **two** uses of ATP in a cell. **[1]**

(Total 6 marks)

(Edexcel 6041, B/HB1, June 1999, Q.4)

Topic ❷

Regulation of the internal environment

Introduction

This part of the unit deals with two aspects of regulation: the maintenance of a constant **internal environment** and the responses to changes in the external environment. In Unit 2, you learnt that living organisms exchange materials, including respiratory gases and excretory products, with their external environment. Here, we look at ways in which organisms control their internal environment, with particular reference to the regulation of blood glucose and the role of the kidney in the control of body water and in the elimination of nitrogenous waste. Homeostasis and the roles of feedback mechanisms are important concepts. An understanding of transport across membranes from Unit 2 will help in the study of this topic.

All living organisms must be able to respond to changes in their external environment and, in order to accomplish this, they need to be able to detect external stimuli before an appropriate response can be made. In this topic, the detection of light by both flowering plants and mammals is studied. The coordination of responses in animals is achieved by means of **hormones** and the **nervous system**. This topic develops your understanding of the principles of hormonal action and control, through a study of the regulation of the blood glucose level by insulin and glucagon, antidiuretic hormone and the reproductive hormones. In addition, you need to understand the role of the nervous system in coordination.

This topic is quite long and requires the study of a variety of different body systems, so it would be sensible to learn each system separately first, so that you can gain an understanding of the basic principles. You will need to bring all your knowledge together finally in order to be able to appreciate the general roles of feedback mechanisms and the differences between nervous and hormonal coordination. Do not forget the brain; you need to know the gross structure and the roles of the major regions.

The required practical work involves experiments on reaction times and a knowledge of the histology of the spinal cord.

Checklist of things to know and understand

Before attempting to answer any of the questions, check that you know and understand the following:

❏ the concept of homeostasis and its importance in maintaining the body in a state of dynamic equilibrium

❏ that homeostasis allows organisms to be independent of the external environment

❏ the concept and roles of feedback mechanisms

❏ the role of the mammalian kidney in osmoregulation and nitrogenous excretion

❏ the structure of the mammalian kidney

❏ the function of the nephrons

❏ that urea is formed in the liver from excess amino acids

❏ that urine formation involves the processes of ultrafiltration and selective reabsorption

❏ the counter-current multiplier

❏ that water and solute content of the blood is controlled by the action of osmoreceptors in the hypothalamus and by antidiuretic hormone (ADH)

❏ the factors which lead to variation in blood glucose levels

❏ the roles of insulin, glucagon and adrenaline in the control of blood glucose levels

❏ the role of the liver in glucose-glycogen metabolism

❏ that organisms need to detect external stimuli

❏ the concept of sensory receptors, illustrated by the detection of light in flowering plants by phytochrome and by retinal pigments in the mammalian eye

❏ the principles of hormonal action and control illustrated by insulin and glucagon in the regulation of blood glucose level, antidiuretic hormone and reproductive hormones, and the principle of negative feedback

❏ the differences between nervous and hormonal coordination

❏ the structure and functions of sensory, relay and effector neurones

❏ the role of Schwann cells and myelination

❏ the nature of the nerve impulse, including changes in the permeability of the membrane to sodium ions, resulting in a wave of depolarisation

❏ the structure and functions of a synapse, the role of acetylcholine (ACh) as a transmitter substance and the generation of postsynaptic potentials

❏ the effects of nicotine on synaptic transmission

❏ the gross structure of the brain and spinal cord

❏ the location and functions of the medulla, cerebellum, hypothalamus and cerebral hemispheres

❏ the functioning of a spinal reflex arc and the types of neurone involved

❏ the value of such reflexes in response to changes in the external environment.

Practicals

You are expected to have carried out practical work to investigate:

❏ *reaction times*

❏ *the histology of the spinal cord.*

Practical work – Helpful hints

Reaction time to a visual stimulus can be measured simply using the 'ruler-drop' test.

● A metre rule is held between the thumb and first finger of a subject and dropped without warning.

● The subject catches the ruler and the distance the ruler has dropped is noted by recording the distance just above the subject's index finger.

● Distance may be converted to time, using the formula below.

$$t = \sqrt{\frac{2s}{g}}$$

where t = time (in seconds)

s = distance dropped (in metres)

g = 9.81 (acceleration due to gravity)

You are also expected to observe microscope slides of transverse sections through the spinal cord.

● When viewing microscope slides, always use the low magnification objective lens (usually × 4) first, before changing to the next magnification objective.

● Use diagrams and photomicrographs in text books to help you interpret the section.

● Make a labelled low power plan of the section, to show the grey matter, the central canal and the white matter.

● You may be able to see cell bodies of effector (motor) neurones in the grey matter.

Helpful hints

The kidney has other homeostatic functions, including regulation of the acid-base balance of the body, and endocrine functions.

It will assist your revision of structure and function of the kidney if you annotate your diagram to show where processes such as ultrafiltration and selective reabsorption occur.

Many different substances are reabsorbed in the proximal convoluted tubule, including sodium ions, calcium ions, sulphate ions, magnesium ions and uric acid. You need not try to remember them all!

Remember that insulin *promotes* the uptake of glucose into body cells, particularly liver and muscle, where the glucose is converted to glycogen. Avoid the temptation to say that insulin converts glucose to glycogen, as this is not strictly true.

 ## Testing your knowledge and understanding

To test your knowledge and understanding of regulation of the internal environment, try answering the following questions.

2.1 Explain what is meant by the term homeostasis.

2.2 Give *two* functions of the kidney.

2.3 Draw and label a diagram to show the structure of a nephron.

2.4 Which component of plasma will not be present in the glomerular filtrate? Explain your answer.

2.5 Name *one* substance which is normally completely reabsorbed from the filtrate in the proximal convoluted tubule.

2.6 Name *two* substances, other than glucose which are reabsorbed in the proximal convoluted tubule.

2.7 Outline the function of the loop of Henlé.

2.8 Suggest how the nephrons of desert mammals are adapted to help conserve water.

2.9 On a hot day, would you expect your circulating concentration of ADH to be high or low? Explain your answer.

2.10 What effect would you expect a diet rich in protein to have on the concentration of urea in urine? Explain your answer.

2.11 Explain why the blood glucose concentration tends to increase after eating breakfast.

2.12 Outline the mechanism which counteracts an increase in blood glucose.

2.13 Name *two* hormones other than insulin which are involved with the regulation of blood glucose and, in each case, state whether the hormone increases or decreases the blood glucose concentration.

You may have read about other hormones which also affect blood glucose, such as glucocorticoids, which increase blood glucose, and growth hormone, which also tends to increase blood glucose concentration.

You could show these changes as a graph, showing changes in the membrane potential against time, during the passage of a nerve impulse. It would be helpful to annotate your graph, to show changes in permeability and the movement of sodium and potassium ions. Remember that sodium ions flood in, followed by potassium ions *out*.

When giving a function of the medulla oblongata, for example, it might be tempting to write 'swallowing' or 'breathing' but this is not really correct! It would be better to say *controls* swallowing, or *controls* breathing.

This is an example of a fill in the gaps question. It is a good idea to read the whole passage through carefully and consider your answers, before filling in the gaps!

2.14 Name *three* pigments which are involved with the detection of light in living organisms.

2.15 Make a table to compare the actions of hormones with the nervous system.

2.16 Compare the effects of *peptide* and *steroid* hormones on a target cell.

2.17 Make a labelled diagram to show the structure of a typical neurone.

2.18 Explain what is meant by the terms *resting potential* and *depolarisation*.

2.19 Make a labelled diagram to show the structure of a synapse.

2.20 List the sequence of events which occur when a nerve impulse arrives at a presynaptic terminal.

2.21 Name *three* chemical transmitter substances.

2.22 Make a table to show *one* function of each of the following parts of the brain: cerebral hemispheres, cerebellum, hypothalamus, and medulla oblongata.

2.23 Explain what is meant by the term *spinal reflex*.

 Practice questions

Mark allocations are given for each part of the questions and the answers are given on pages 107–8.

1 Read through the following account of kidney function, then write on the dotted lines the most appropriate word or words to complete the account.

> In the kidney, the renal artery branches to form many smaller arterioles, each of which divides further to form a knot of capillaries called a .... Here, small molecules such as .... and .... are forced into the cavity of the Bowman's (renal) capsule by the process of .... .Selective reabsorption takes place in the nephron. In the proximal convoluted tubule all the .... is reabsorbed. In the ascending limb of the loop of Henlé, ions are pumped out of the nephron. This causes .... to be drawn out of the collecting duct.

(Total 6 marks)
(Edexcel 6043, B3, January 1998, Q.2)

2 In the table below, columns 1 and 2 show the quantities of water, glucose and urea passing through Bowman's capsules and the collecting ducts of a kidney in a 24 hour period. Columns 3 and 4 show the quantities and percentages reabsorbed during the same period.

Complete the table by writing the correct figures in the boxes labelled (i) to (iv).

Substance	Quantity passing through Bowman's capsules	Quantity passing through collecting ducts	Quantity reabsorbed	Percentage reabsorbed
Water	180 dm³	1.5 dm³	178.5 dm³	(i)
Glucose	180 g	(ii)	180 g	100
Urea	53 g	25 g	(iii)	(iv)

(Total 4 marks)
(Adapted from Edexcel 6043, B3, June 1996, Q.4)

3 Read through the following passage, which refers to the detection of light in the mammalian eye, and then write on the dotted lines the most appropriate word or words to complete the passage.

Light is detected by specialised cells in the retina of the eye. The rod cells detect light at .... intensity. They contain a photosensitive pigment called .... , which consists of a protein molecule, ... , joined to a molecule of retinal. Light causes the pigment to separate into its two component molecules. When we move from the light to the dark the pigment is re-synthesised. This process is known as .... adaptation.

(Total 4 marks)

(Edexcel 6043, B3, June 1998, Q.4)

4 Explain what is meant by each of the following terms.

   (a) Action potential. **[2]**

   (b) Transmitter substance. **[2]**

   (c) Myelination. **[2]**

(Total 6 marks)

(Edexcel 6043, B3, January 1997, Q.5)

> ### Helpful hints
>
> Note that there are two marks allocated for each part of this question, so the Examiner will expect two relevant facts in each case.

5 The diagram below shows a human brain seen from the right side.

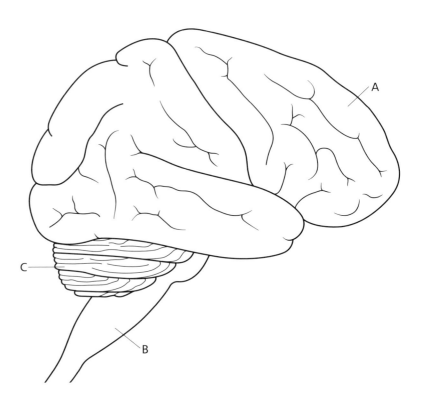

   (a) Name the parts labelled A, B and C. **[3]**

   (b) Give **two** functions of the part labelled B. **[2]**

(Total 5 marks)

(Edexcel 6043, B3, June 1997, Q.1)

Helpful hints

This question is to test your ability to interpret data from an experiment, presented in the form of a graph, and your knowledge and understanding of the control mechanisms of blood glucose concentration. When looking at the graph, note carefully the units which are units for the glucose concentrations and for the insulin concentrations.

6 An experiment was carried out to investigate the relationship between concentration of glucose and insulin in the blood of healthy people. At the start of the experiment 34 volunteers each ingested a syrup containing 50 g of glucose. The concentration of glucose and insulin was determined in blood samples at intervals over a period of 2 hours.

The results are shown in the graph below and are mean values for the group of volunteers.

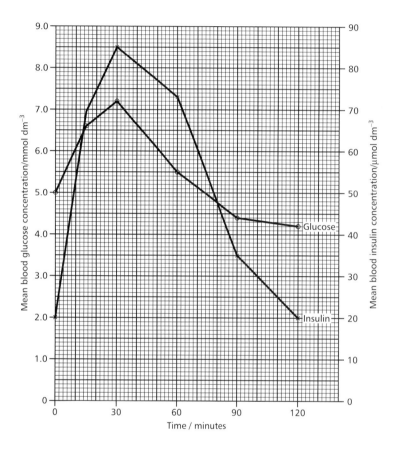

(a) From the graph, find the mean concentration of insulin 100 minutes after the start of the experiment. **[1]**

(b) Describe and suggest reasons for the changes in the concentration of glucose during the following time intervals.

   (i) 0 to 30 minutes. **[2]**

   (ii) 30 to 120 minutes. **[2]**

(c) Describe and explain the relationship between the concentrations of glucose and insulin as shown by this graph. **[3]**

(d) Name **one** hormone, other than insulin, which is involved with the regulation of blood glucose and state is effect on blood glucose concentration. **[2]**

(Total 10 marks)
(Edexcel 6046, B6, January 1996, Q.2)

Mark allocations are given for each part of the questions and the answers are given on page 124.

Unit 4

Core section: Assessment questions

1   The diagram below summarises the stages of respiration in muscle tissue.

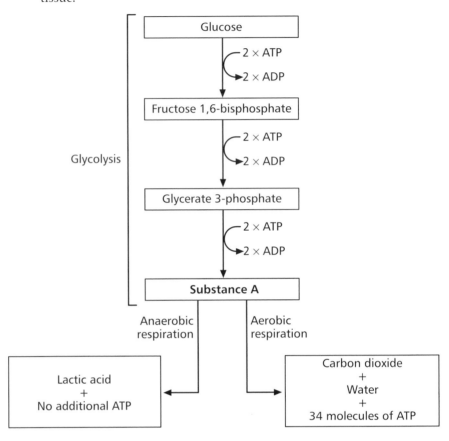

(a) Name Substance A.   **[1]**

(b) What is the net yield of ATP produced by the breakdown of one molecule of glucose under anaerobic conditions?   **[1]**

(c) Using the information in the diagram, calculate the yield of ATP produced by anaerobic respiration as a percentage of that produced by aerobic respiration. Show your working.   **[2]**

(Total 4 marks)
(Edexcel 6041, B/HB1, June 2001, Q. 3)

2   The diagram on the left shows a rod cell from the retina of a mammal.

(a) Give **one** function for each of the parts labelled A and B on the diagram.   **[2]**

(b) Explain why an image focused on to the fovea appears more detailed than an imaged focused on to another part of the retina.   **[3]**

(Total 5 marks)
(Edexcel 6043, B3, June 2001, Q.4)

3 The diagram below represents a kidney tubule (nephron).

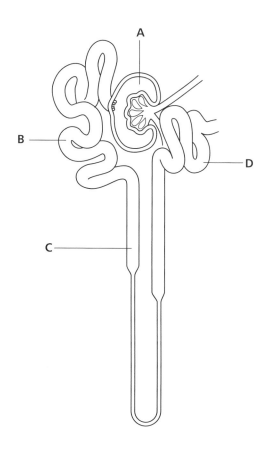

(a) Name the parts labelled A, B, C and D. **[2]**

(b) The concentration of urea was measured in parts of the nephron labelled A, B and D.

Part of nephron	Concentration of urea/g dm^{-3}
A	0.30
B	0.55
D	6.0 – 7.0

  (i) Explain how the change in the urea concentration between part A and part B is brought about. **[2]**

  (ii) Explain why the urea concentration in part D of the nephron may vary at different times of the day. **[3]**

(c) When this person ate a high protein diet, the urea concentration in part A of the nephron was 0.36g dm^{-3}.

  (i) Calculate the percentage change in the urea concentration in part A resulting from this change in diet. **[3]**

  (ii) Give an explanation for this change in urea concentration. **[3]**

(Total 13 marks)
(Edexcel 6043, B3, June 2001, Q.7)

4   Explain the meaning of the following terms.

    (a) Action potential.   [3]

    (b) Myelination.   [3]

(Total 6 marks)
(Edexcel 6043, B3, January 2001, Q.3)

5   Give an account of the structure of mitochondria and their role in respiration.

(Total 10 marks)
(New question)

The Options

As part of Unit 4, students are requested to select one option from the following list:

A Microbiology and biotechnology
B Food science
C Human health and fitness.

Each option relies on some knowledge of topics studied in the core units and provides students with an opportunity to specialise and extend their biological studies in a selected area. In each option, there is an emphasis on the applications and relevance of biology in everyday life.

For each option you will find a limited number of questions or activities to test your knowledge and understanding. These provide you with a starting point for your revision and you can also devise your own ways of testing yourself, using examples from the core units. The strategies for effective revision apply here, as they do to all the sections of the specification, and the range of activities listed in the Activities toolkit will also help you.

For each option, you will find:

- an **introduction**
- a comprehensive **checklist of things to know and understand**
- a **checklist of required practicals**
- **helpful hints** for practical work
- some **questions** and **activities** to test your knowledge and understanding
- some **assessment questions**.

Practical work is an important component of each option and it is wise to ensure that you are familiar with the details of all the experiments and investigations described in your chosen option.

Where appropriate, special hints relevant to a specific option have been included.

Option Ⓐ Microbiology and biotechnology

 Introduction

This option is divided into three sections:

A1 Diversity of microorganisms
A2 Culture techniques
A3 Use of microorganisms in biotechnology

It is clear that there is an emphasis on practical work and you should be familiar with all the techniques described and the investigations listed at the end of each section.

You will find that review and revision of core topics such as molecules (Unit 1.1), cellular organisation (Unit 1.3), modes of nutrition (Unit 3.1) and aerobic and anaerobic respiration (Unit 4.1) will help you. You should note that much of this material is in the AS part of the specification, so you may need to go back to the notes and revision cards that you used before.

The checklist below has been subdivided according to the sections in the option. It may be sensible to do your revision section by section, rather than attempting to revise the whole option in one go.

 Checklist of things to know and understand

Before attempting any of the questions or activities, check that you know and understand the following:

A1 Diversity of microorganisms

❏ the general characteristics of bacteria and fungi (yeasts and moulds)

❏ the structure of a bacterial cell and its inclusions, illustrated by *Escherichia coli*

❏ the importance of cell structure as a means of classification of bacteria

❏ the use of Gram staining in identifying bacteria

❏ bacteria as agents of infection; the production of exotoxins (*Staphylococcus*) and endotoxins (*Salmonella*); the invasion of host tissue (*Mycobacterium tuberculosis*)

❏ the differences in structure between yeasts and moulds, illustrated by *Saccharomyces* and *Penicillium*

❏ the classification of viruses based on nucleic acid types as illustrated by λ (lambda) phage (DNA), tobacco mosaic virus (RNA) and human immunodeficiency virus (RNA retrovirus)

❏ viruses as agents of infection; the nature of host cell specificity; cell infection cycle and latency as illustrated by human immunodeficiency virus (HIV)

A2 Culture techniques

- ❏ the essential nutrients (carbon sources, nitrogen sources, minerals and growth factors) appropriate to the growth of heterotrophic microorganisms

- ❏ the environmental influences of temperature, oxygen level and pH on growth

- ❏ the principles and techniques involved in culturing microorganisms

- ❏ the use of different media (solid and liquid media, selective media, indicator media)

- ❏ the use of fermenters (bioreactors) for the production of mycoprotein and antibiotics

- ❏ the differences between batch and continuous fermentation

- ❏ that industrial processes involve the need for aseptic entry of material, culture inoculants, media, aeration, temperature, pH, agitation, product recovery and how culture conditions may be controlled

- ❏ the stages of growth of microorganisms in culture; diauxic growth; the production of secondary metabolites

- ❏ methods of measuring culture growth as illustrated by cell counts, dilution plating, mass and optical methods (turbidity)

- ❏ the construction of growth curves and the calculation of growth rate constants

A3 Use of microorganisms in biotechnology

- ❏ the processes involved in lactic acid fermentation leading to the production of yoghurt

- ❏ the processes involved in fermentation by yeast in brewing and dough production and the metabolic processes involved

- ❏ the production of mycoprotein

- ❏ the production of antibiotics, illustrated by penicillin from *Penicillium*

- ❏ the effects of antibiotics (penicillin) on bacterial growth

- ❏ antibiotic resistance and the reasons for its spread.

Practicals

You are expected to have carried out the following practical work to investigate:

- ❏ *the use of Gram staining in the identification of bacteria*

- ❏ *the preparation and sterilisation of media, agar-plate pouring and inoculation using sterile wire loops, pipettes and spreaders*

- ❏ *the use of different carbon and nitrogen sources for growth using cultures on agar plates or in liquid culture*

- ❏ *the use of different methods for measuring the growth of a suitable microorganism in a liquid culture*

- ❏ *the optimal conditions (e.g. temperature, pH, nutrients) necessary for yoghurt production or for fermentation by yeast in brewing or dough production.*

〴〳〲 *Practical work – Helpful hints*

The Gram staining method is an important technique for the identification of bacteria, which are divided into two groups, known as Gram positive and Gram negative. Gram positive bacteria retain a crystal violet complex when treated with organic solvents, such as ethanol, and appear purple. Gram negative bacteria are decolourised by the organic solvents and are then counterstained using, for example, safranin. Gram negative bacteria appear pink or red.

- Prepare a heat-fixed smear of bacteria on a clean microscope slide.
- Stain the smear with crystal violet. Leave the stain for 30 seconds, then rinse off with tap water. Flood the slide with iodine solution. Leave for a further 30 seconds, then rinse the slide with tap water.
- Rinse the slide with alcohol, until the washings are pale violet, but be careful not to over-decolourise. Rinse again with tap water.
- Counterstain with safranin for one minute, rinse the slide with tap water, then blot dry.
- Examine the slide using an oil-immersion objective.
- Remember to wipe the oil off the lens after use.
- Examples of Gram positive bacteria include *Bacillus subtilis* and *Streptococcus thermophilus*.
- Examples of Gram negative bacteria include *Escherichia coli* and *Salmonella enteritidis*.

This section also introduces you to aseptic technique for pouring agar plates and using sterile wire loops, pipettes and glass spreaders. You are also expected to carry out investigations into the effects of different carbon and nitrogen sources on the growth of microorganisms.

- Microbiological media are usually sterilised in an autoclave, which uses steam under pressure to achieve a high temperature, such as 121 °C.
- After sterilisation, allow the media to cool to about 50 °C before pouring into sterile Petri dishes, using aseptic technique.
- Sterile pipettes are used to measure volumes of, for example, broth cultures, or when carrying out dilution plating.
- Wire loops are sterilised by introducing into a Bunsen flame until the loop is red hot. Wire loops are used for making streak plates.
- Glass spreaders are usually sterilised by dipping into alcohol, then passing through a Bunsen flame and allowing the alcohol to burn off. Glass spreaders are used for spreading a measured volume of culture over the surface of an agar plate to produce uniform growth.
- To investigate the effect of different carbon sources on growth, yeast can be cultured in media containing a range of sugars, such as glucose, fructose, galactose, sucrose and maltose.
- To investigate the effect of different nitrogen sources on growth, you could try culturing yeast in media containing ammonium sulphate, amino acids or potassium nitrate.
- Remember that good experimental design is essential. This includes changing only one variable and keeping all other factors constant.

A further investigation involves growing a microorganism, such as yeast, or *Chlorella*, in a liquid medium. The organism could be grown in a fermenter, or bioreactor, and samples removed at regular time intervals.

- Set up the culture and remove samples at suitable time intervals.
- Determine the number of cells using direct counting, dilution plating or optical methods.
- Direct counting involves the use of a haemocytometer, or counting chamber. This gives the total number of cells present, including both viable and non-viable cells.

- Dilution plating gives the number of viable cells, and relies on the principle that viable cells will grow to produce visible colonies on the surface of an agar plate.
- Optical methods involve the use of a colorimeter and are particularly useful for investigating the growth of an organism such as *Chlorella*.
- You could use a combination of the methods above and compare results.
- Plot a growth curve as number of cells (or absorbance) against time.
- You may be able to calculate the exponential growth rate constant (*k*) from your results.

You are also expected to investigate the optimal conditions, such as temperature, pH or nutrients, necessary for yoghurt production or for fermentation by yeast in brewing or dough production.

- You could investigate yoghurt production by determining changes in viscosity or changes in pH.
- Add 1.0 cm³ of natural yoghurt, as a starter culture, to 10.0 cm³ of milk in a boiling tube, mix, and cover with cling wrap.
- Incubate in a water bath at 43 °C.
- Observe changes in viscosity and measure changes in pH
- You could investigate the effects of using different types of milk, such as pasteurised, UHT, and lactose-reduced milk.
- To investigate fermentation by yeast in brewing, you could compare the growth of *Saccharomyces cerevisiae* and *S. carlsbergensis* in media containing different sugars, such as glucose, sucrose and maltose.
- To investigate factors affecting dough production, you could devise methods to compare the effects of factors including different strains of yeast, types of flour and the concentration of ascorbic acid, on the rate of rising of dough.
- To prepare dough, mix 1 g of dried yeast with 50 cm³ of water, then add 75 g of flour and mix well.
- Place the dough in a 100 cm³ measuring cylinder and record the volume of the dough at suitable intervals.

Helpful hints

You could try to organise this information into a series of headings so that you are familiar with the generalised cell structure of bacteria and of the fungi you study, and compare each of these with each other and with a 'typical' plant (or animal) cell. You can then look at the characteristics of viruses and see why they are put into a separate group. A good way to test yourself is to draw diagrams of a 'typical' (or *named*) bacterium, fungus and virus and then to annotate these diagrams with particular features that are significant when you are considering the diversity of these organisms.

 Testing your knowledge and understanding

A1 Diversity of microorganisms

A1.1 Start yourself off with a brainstorming session about microorganisms. Use the ideas below to help organise your information relating to microorganisms and their diversity. First work out *why* we put microorganisms into groups, separated from the plants and animals that you are more familiar with. Then think *how* 'microorganisms' differ from plants and animals that you have studied. Is it easy just to talk about a 'typical' microorganism? List some *activities* that microorganisms carry out and think why these are important to the living world on a broad scale and to humans in particular.

Unit 4

Helpful hints

You should do this first in relation to your own practical techniques for small scale work in the laboratory, then adapt the same principles for the systems used for growing microorganisms in huge bioreactors on an industrial scale.

Make sure you refer to how you prepare the media, pour the plate and mention how you use wire loops, pipettes and spreaders. It is important that you give details of appropriate sterile techniques and the conditions you would use to grow your culture after inoculation.

Helpful hints

Follow through your laboratory flow chart (summarising *your* techniques on a small scale) and your simplified (but annotated and improved!) diagram of a bioreactor (representing large scale operations). Make sure that you identify conditions for entry of the different ingredients (including microorganism), the product recovery and how culture conditions (such as temperature, pH and aeration) are controlled.

A2 Culture techniques

A2.1 Next look at the **techniques** required for growing microorganisms.

(a) Draw a flow chart to help you summarise the steps you would take to prepare and grow a culture on an agar plate.

(b) Now extend the same principles to growth of microorganisms in liquid culture in a bioreactor on an industrial scale. Start with the box (below) to represent the bioreactor. Annotate and add to the diagram to show the inputs, how (and when) the product is harvested and how the required conditions are maintained during the growth of the microorganism.

> **industrial bioreactor**
> *add inputs, outputs,*
> *how correct conditions*
> *are maintained*

(c) Finally link this bioreactor to industrial fermenters (bioreactors) used in the production of **mycoprotein** and **antibiotics**.

A3 Use of microorganisms in biotechnology

A3.1 Brewing involves a fermentation by yeast of sugars, usually using barley grains as the source material. There are several stages in this process, as listed below.

List of stages in brewing *– fermentation ; mashing ; malting ; milling ;*

(a) Select the correct stage from the list and put this onto the dotted lines labelled [A1], [A2] etc. This then gives the sequence of these stages in the brewing process. Next find a suitable word or words to complete the outline of essential events during that stage. Lastly give examples of the enzyme reactions occurring during that stage or which reflect the metabolic or other important changes taking place

[A1] . . . . Barley . . . . are . . . . in water and allowed to . . . . The . . . . are then are spread out on the . . . . and turned at intervals. The growth substance . . . . acts on the . . . . layer in the grain and stimulates release of certain enzymes. Germination is stopped by kilning (heating to between . . . . and 80 °C).

Examples of enzyme reactions: . . . .

[A2] . . . . The . . . . are crushed so that the . . . . endosperm is broken into grist, which is a . . . . flour.

[A3] . . . . Grist is mixed with . . . . in a . . . . and enzyme activity converts starch to soluble . . . . This produces a sweet liquid, known as . . . .

Examples of enzyme reactions: . . . .

[A4] . . . . Addition of yeast allows the sweet liquid (. . . . .) to ferment. The yeast . . . . is generally used in the production of ales and the yeast . . . . is used in the production of lager.

Examples of comments: . . . .

(b) Why are the grains turned during the malting process?

(c) When is sugar *added* to the brewing liquid?

Helpful hints

Helpful hints: To do your revision on yoghurt and bread making you can refer to a series of questions on the topic of fermentation in the *Food Science* section on page 30. Work through these tables and questions and select those which are relevant (or adapt them) for yoghurt and for bread. (To give you a start with yoghurt, see Food Science, Qu. B4.1.)

(d) When is the brewing liquid boiled with hops? What does the boiling achieve? What is the name of the vessel used at this stage?

(e) Give the temperatures and approximate times for the fermentation stage, for ales and for lagers.

(f) When the fermentation is complete, what treatments are needed before the ale or lager is ready for sale and consumption?

(g) Name *three* byproducts from the brewing process. At which stage is each produced? What can they be used for?

 Special hints for revision

Besides revising each major section of the option, it could be helpful to link subjects that appear in different sections. For example, the structure of *Penicillium* (A1) could be linked to the production of antibiotics (A2 and A3) and to their effects on bacterial growth (A.3). There are similar links with yeasts and fermentation processes (A1, A2 and A3). Short-answer questions often require definitions, naming structures on diagrams or precise practical details, so it is important to be thoroughly familiar with the material in the option, so that you are confident that you can answer these straightforward questions easily. This will leave you more time to think about those questions where you may be confronted with unfamiliar data and where you need to apply your biological knowledge.

Assessment questions

Mark allocations are given for each part of the questions and the answers are given on pages 124–5.

1 The graph below shows the number of new cases of HIV infections and AIDS in England each year between 1992 and 1998.

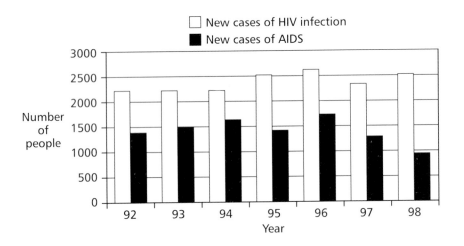

(a) The graph shows that the number of new AIDS cases reported each year has decreased between 1996 and 1998.

    (i) Calculate the percentage decrease in the number of AIDS cases reported between 1996 and 1998. Show your working   **[2]**

    (ii) Suggest **one** reason for this change.   **[2]**

(b) Explain why the number of people known to be infected with HIV is always likely to be greater than the number of AIDS cases reported.   **[3]**

(c) Suggest why the number of people actually infected with HIV may be much larger than the data suggest.   **[2]**

(d) Explain why HIV can be described as a **retrovirus**.   **[2]**

(Total 11 marks)
(Edexcel 6044/01, B/HB4A, June 2001, Q.7)

2   The diagram below shows some stages in the dilution and plating method for estimating the number of bacteria in a sample of milk.

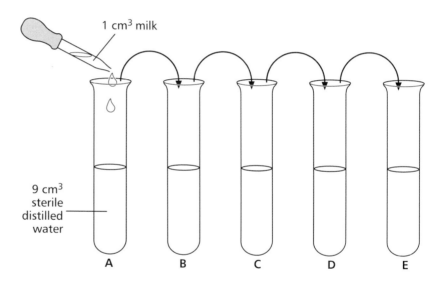

1 cm³ milk

9 cm³ sterile distilled water

A   B   C   D   E

1 cm³ of milk was pipetted into 9 cm³ of sterile distilled water in tube A and mixed thoroughly. 1 cm³ from tube A was then pipetted into tube B and the same procedure carried out until five dilutions had been made. 0.1 cm³ samples from each of the test tubes A, B, C, D and E were plated out and incubated for 2 days.

(a) Explain why dilution of the original sample may be necessary. [2]

(b) Describe the technique used for plating out the samples. [3]

(c) The diagrams below show the bacterial colonies on plates made from C, D and E.

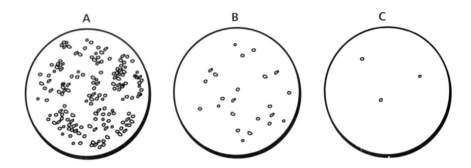

(i) State which plate you would use to count the number of colonies in order to estimate the number of bacteria in the original milk sample. Give reasons for your answer. [3]

(ii) Calculate the number of bacterial cells in 1 cm³ of the original milk sample. Show your working. [3]

(d) State **one** reason for the use of this method of estimating numbers of microorganisms rather than the use of a direct counting method such as a haemocytometer. [1]

(Total 12 marks)

(Edexcel 6044/01, B/HB4A, June 2000, Q.6)

3 In the culturing of microorganisms in the laboratory it is necessary to use aseptic techniques to prevent contamination.

The diagram below shows how a simple autoclave is used to sterilise laboratory equipment and culture media.

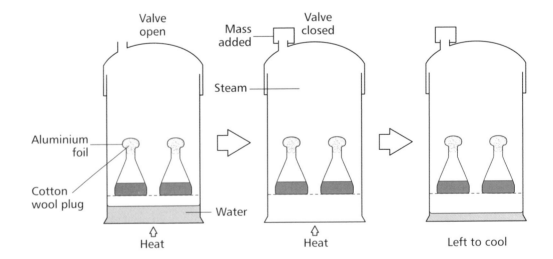

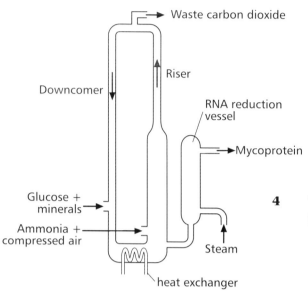

Waste carbon dioxide

Riser

Downcomer

RNA reduction vessel

Mycoprotein

Glucose + minerals

Ammonia + compressed air

Steam

heat exchanger

(a) The pressure of steam builds up to 103 kPa. This raised the boiling point of water to 121 °C. The contents of the autoclave are maintained at this temperature for 15–20 minutes.

Explain why these conditions are necessary for effective sterilisation. **[2]**

(b) Suggest why the cotton wool plugs of the flasks are covered with aluminium foil. **[1]**

(c) Describe and explain **two** precautions, other than the use of sterile equipment, you would need to take to prevent contamination of cultures of microorganisms. **[4]**

(Total 7 marks)

(Edexcel1 6044/01, B/HB4A, January 2001, Q.5)

4 The diagram shows an air-lift fermenter which can be used in the production of mycoprotein.

(a) Name the fungus used in the production of mycoprotein. **[1]**

(b) Explain why ammonia and compressed air are added to the fermenter. **[3]**

(c) The RNA reduction vessel reduces the RNA content of the mycoprotein to levels suitable for human consumption. Suggest why steam is added to this vessel. **[2]**

(d) The heat exchanger consists of a coil through which cold water is passed. Suggest the reason for the heat exchanger in the fermenter. **[2]**

(Total 8 marks)

(Edexcel 6044/01, B/HB4A, June 2001, Q.5)

Option B

Food science

 Introduction

This option is divided into four sections:

B1 Food and diet
B2 Food additives
B3 Food storage
B4 Biotechnology and food production

All the sections involve some practical work, with part of the emphasis on the requirements of the human diet and how these can be met. There is some exploration of diseases resulting from poor diets. Applied aspects consider the biology of the changes in fruits and vegetables after harvest and in storage. It also looks at some fermentations as a means of modifying foods.

You will find that review and revision of core topics such as molecules (Unit 1.1), digestion and absorption (Units 2B.1 and 2H.1), movement of water (Unit 2B.2) and respiration (Unit 4) could be helpful. Some of the topics appear to have few links with core material, but there are links within the Option which are significant. Many of the linked Core topics are in the AS specification, so you may need to look back at your notes from last year.

The checklist has been subdivided according to the sections in the option and you are recommended to tackle one section at a time rather than trying to revise all of the material in one go. It would be sensible to complete your revision of the whole option before attempting any of the questions.

Checklist of things to know and understand

Before attempting any of the questions or activities, check that you know and understand the following:

B1 Food and diet

❏ the sources and roles of nutrients required in a balanced diet, including carbohydrates, fats, proteins, vitamins, mineral ions, dietary fibre and water

❏ the nutritional requirements with reference to energy, total fat, polyunsaturated and saturated fat, dietary fibre, sodium and sugars

❏ variations in energy requirements in relation to basal metabolic rate, lean body mass, thermogenesis and different physical activities including exercise

❏ the effects of lack of protein, iron, calcium, vitamin C (ascorbic acid) and vitamin A (retinol)

❏ the molecular basis of scurvy in terms of the hydroxylation of collagen

❏ the absorption of iron is influenced by its source (haem or non-haem), the presence of inhibitors or enhancers and the amount already stored in the body

❏ protein-calorie malnutrition, anorexia nervosa and bulimia

❏ the causes and effects of overweight and obesity

❏ calculations of body mass index (BMI)

❏ the possible relationships between diet and coronary heart disease, diseases of the colon and mature onset diabetes mellitus

❏ the relationship between restricted energy intake, physical activity and weight loss; the dangers of very low-calorie and restricted diets

B2 Food additives

❏ the reasons for the use of natural and artificial sweeteners in food processing, illustrated by sucrose and aspartame

❏ comparison of the relative sweetness of different naturally occurring sugars (sucrose, fructose, glucose)

❏ the use of the enzymes glucose isomerase and amyloglucosidase in the processing of food

❏ the nature of lactose intolerance and the use of lactose-reduced milk

❏ the reasons for the use of colourings; distinguish between the use of natural colourings (β-carotene) and artificial colourings (sunset yellow, tartrazine)

❏ the problems associated with the use of artificial colourings

❏ the reasons for the use of antioxidants in the preservation of food, illustrated by ascorbic acid and tocopherol

❏ the distinction between the use of flavourings, such as vanilla, and flavour enhancers, such as salt and monosodium glutamate

❏ the reasons for the use of preservatives, such as sulphites, and the possible problems associated with their use

B3 Food storage

❏ that metabolic processes continue after harvesting

❏ the ripening and development of sweetness in apples

❏ colour development and softening of tissues in tomatoes

❏ the effects of continued respiration and loss of water in fruits and vegetables

❏ the principles of storage in relation to maintaining quality and avoiding spoilage by microorganisms

❏ factors which affect the growth and multiplication of microorganisms in food

❏ the effects of modified atmosphere storage on respiration and delay in ripening

❏ the differences between short- and long-term storage, illustrated by the pasteurisation and sterilisation of milk

❏ the reasons for the choice of packaging materials

❏ the use of plastic films, shrink packs, vacuum packs

❏ the modification of the atmosphere within the pack

B4 Biotechnology and food production

❏ microorganisms (bacteria and fungi) can be used to modify foods

❏ the conversion of raw cabbage to sauerkraut, milk to yoghurt and soya beans to tofu and soy sauce

❏ the changes in pH in relation to storage

❏ the role of yeast in bread-making

❏ the effect of ascorbic acid on the rising of dough

❏ the role of yeast in wine-making.

Practicals

You are expected to have carried out the following practical investigations:

❏ *the determination of the calorific values of simple foods using a calorimeter*

❏ *the estimation of subcutaneous fat by skinfold measurements*

❏ *the perception of sweetness in drinks or food*

❏ *the quantitative estimation of sugars and ascorbic acid at various stages of storage*

❏ *weight loss in packaged foods*

❏ *the resazurin test, methylene blue test and turbidity test in relation to milk of different ages and the effectiveness of pasteurisation and sterilisation*

❏ *the changes in foods during the process of fermentation.*

Practical work – Helpful hints

In this section, you are expected to determine the calorific values of foods using a calorimeter and to estimate subcutaneous fat using skinfold calipers.

● To determine the calorific value of food, a small sample is burnt in a heat of combustion apparatus, or food calorimeter. This provides a means of working out the heat produced by the burning food.

● The principle behind this method is that the heat produced increases the temperature of water in the calorimeter.

● One calorie is the amount of heat required to raise 1 g of water by 1 °C.

● The heat produced by the food is therefore the mass of water (in g) × the rise in temperature in °C.

● To convert calories to joules, multiply by 4.2 (1 calorie = 4.2 joules).

● Skinfold measurements provide a method for estimating the percentage of body fat.

● Skinfold calipers have a spring which exerts a standard pressure and a scale to measure skinfold thickness in millimetres. Measurements of skinfold thickness are taken at four sites, at the front of the upper arm, the back of the upper arm, on the back just below the shoulder blade, and on the side of the waist.

● These readings are added together, then Tables are used to convert to a percentage body fat.

● A less accurate estimation can be obtained using skinfold measurements from the back of the upper arm only. Again, these measurements are converted to percentage body fat using Tables.

The degree of sweet taste of different sugars varies, so some sugars taste sweeter than others. In this investigation, you are expected to compare the relative sweetness of, for example, sucrose, maltose, lactose, glucose and fructose. You could also compare the sweetness of confectionery products – chocolate made for the British market is often much sweeter than that made for other European countries.

● Taste 4 % solutions of sucrose, lactose, maltose, glucose and fructose, rinsing your mouth out with water between each solution.

● Decide which solution tastes the most sweet and which the least sweet.

● Try to arrange the sugars in order of sweetness.

● The degree of sweetness of sugars is usually compared with that of sucrose, which is given a relative sweetness value of 1.0

● Artificial sweeteners include aspartame and saccharin.

Another practical requires you to make quantitative estimations of the sugar and ascorbic acid (vitamin C) content of fruits at various stages of storage.

● Reducing sugars can be estimated using Benedict's reagent and a range of colour standards, made by heating 3.0 cm³ of glucose solutions, of known concentrations, with 5 cm³ of Benedict's reagent.

● The concentration of glucose can be determined using reagent test strips, such as Diabur-Test 5000.

● For non-reducing sugars, it is necessary first to hydrolyse the sugar by treatment with hydrochloric acid, or sucrase, before applying the test for reducing sugars.

● Ascorbic acid concentrations are determined using the DCPIP decolourisation method.

● This involves finding the volume of a standard ascorbic acid solution required to decolourise exactly 1.0 cm³ of DCPIP solution.

● The procedure is repeated using fruit juice and a fresh 1.0 cm³ sample of DCPIP.

Unit 4

- If the volume of standard ascorbic acid solution required to decolourise 1.0 cm³ of DCPIP solution is x cm³ and the volume of fruit juice required to decolourise 1.0 cm³ of DCPIP is y cm³, the concentration of ascorbic acid in the fruit juice is $x \div y$ mg per cm³.

In this section, you are also expected to carry out simple experiments to investigate weight loss in packaged foods, and the resazurin test, the methylene blue test and turbidity test in relation to milk of different ages and the effectiveness of sterilisation and pasteurisation.

- To investigate weight loss in packaged foods, you could investigate the effects of various packaging materials, such as cling wrap, PVC films and paper, on weight loss in various fruits and vegetables. Record your results in a table and plot graphs to show the percentage change in mass against time.

- **Resazurin** is an indicator which shows metabolic activity of bacteria. Resazurin in blue in the oxidised state but, when it is reduced, changes through pink to white.

- This test is used to compare the bacterial content of milk samples. Tubes containing milk and resazurin solution are incubated in a water bath at 37 °C for one hour. Tubes containing milk which change colour to white, pink, or white mottling have failed the test.

- **Methylene blue** is a redox indicator which shows metabolic activity of bacteria in the milk sample.

- Methylene blue changes to colourless when it is reduced, so recording the time taken for the blue colour to disappear gives an indication of bacterial activity in a milk sample.

- The **turbidity test** is used to check the efficiency of sterilisation.

- A milk sample is treated with ammonium sulphate, left for 5 minutes, then filtered. The filtrate is then heated in a boiling water bath for 5 minutes.

- If sterilisation has been effective, the filtrate should remain clear.

Another practical requires you to investigate changes in foods during the process of fermentation. You could, for example, record changes in the pH during the production of yoghurt or changes in pH during the production of sauerkraut.

- Transfer 10 cm³ of UHT milk to a boiling tube and add 1.0 cm³ of natural yoghurt as a starter culture.

- Mix, then record the pH. Cover the tube with cling wrap, then incubate in a water bath at 43 °C.

- Record changes in the pH of the contents at suitable intervals, such as every 30 minutes.

 Testing your knowledge and understanding

B1 Food and diet

B1.1 How far does 'fast-food' affect your or the nation's health?... If you are uncertain, take a look at how numbers of fast-food restaurant have grown (globally) over the last 50 years. Here are some estimates of numbers (world-wide) of fast-food restaurant in the top 10 US fast-food chains. In the 1950s, there were virtually no fast-food restaurants, yet there has been a 10-fold increase from about 10 000 in 1970 to over 100 000 in the year 2000. Now read the passage which makes comments on this trend. You will find it refers to

Unit 4

Helpful hints

Make sure that your answers refer to the *biology* you have learnt and that you think out your responses from a scientific basis. You are not being asked for a value judgement as to whether fast-food restaurants are good or bad!

Helpful hints

Probably the most sensible way to organise your answer is in a table.

several of the 'key' factors related to diet which you study in this topic. Then use the questions below as a guide to your revision.

1 "One reason these trends are worrisome is the health effects of fast food, which is often high in calories, sodium, fat, and cholesterol. At many fast-food restaurants, a single meal gives a disproportionate share – and sometimes more than 100 percent –
5 of the recommended daily intake of these elements. Excessive consumption of fast food produces a diet high in saturated fats and low in fruits and vegetables, which increases the risk of obesity, coronary heart disease, hypertension, diabetes, and several forms of cancer. Fifty-five percent of Americans are
10 considered overweight or obese, the result of poor eating habits (often including excessive consumption of fast food) and a sedentary lifestyle. In developing countries, too, the prevalence of obesity, hypertension, and coronary heart disease are all much greater in urban areas – where fast food and street vendors are
15 commonly found – than in rural areas."

(a) From the passage, pick out the features listed in fast food that could be a cause for worry in the diet. Then for each, say briefly *why* it could be 'worrisome' (line 1).

(b) The passage (line 7) suggests these foods increase the risk of certain conditions. Pick out and make a list of these conditions and say briefly what problems each may bring to the person.

(c) List some factors *other* than those referred to in this passage that may lead to obesity in certain people. Then give some of the conditions (*other* than those already referred to in the passage) which might have increased risk in obese people.

(d) Suggest ways that obesity can be minimised or treated. Make sure you include any precautions that should be observed.

B1.2 Look now at some of the effects of undernutrition.

The list below gives names associated with different nutrition disorders or deficiencies. Select the correct 'disorder' or 'deficiency' from the list and put this onto the dotted lines labelled [A1], [A2] etc. Then find a suitable word or words to complete the descriptions.

List of disorders or deficiencies – *anaemia ; anorexia nervosa ; bulimia ; kwashiorkor ; marasmus ; scurvy ;*
[A1] . . . . is an . . . . disorder in which the person tends to overeat, but this is followed by . . . . This results in some loss of . . . . and other symptoms include . . . .

[A2] . . . . is due to lack of vitamin . . . . (also known as . . . . acid) which is needed for . . . . of collagen. This link with collagen is related to the inability of . . . . to heal successfully. Other symptoms include . . . .

[A3] . . . . is linked with lack of . . . . The person affected has a lack of . . . ., an essential component of . . . . in . . . . blood cells and . . . . in muscle cells. The person affected has a shortage of . . . . cells, often a . . . . appearance and be breathless on exertion.

[A4] . . . . is a . . . . -energy malnutrition which affects young children only. Affected children may suffer severe oedema and have an extremely . . . . belly. The condition may occur after prolonged . . . . when the baby is weaned on to an inadequate traditional diet. Other symptoms include . . . .

[A5] . . . . is an . . . . disorder in which the person may refuse food and has . . . . feeling of hunger. Results are an extreme loss of . . . . and . . . . of BMR.

[A6] . . . . is a . . . .-energy malnutrition in which the food intake is inadequate in relation to . . . . expenditure. Symptoms include severe . . . . of the body and a child may look 'old and . . . .'. The condition can occur in adults as well as children.

B2 Food additives

B2.1 The list below gives four sugars found naturally in foods and one substance used as an 'artificial' sweetener.

List of sugars – *glucose ; fructose ; lactose ; sucrose ; aspartame ;*

(a) Which *two* of these are disaccharides? Name the constituent monosaccharides for each.

(b) Which of these is used as the standard when describing 'relative sweetness'? Arrange these five substances in the order of their relative sweetness, starting with the sweetest.

(c) How would you compare the relative sweetness of glucose, lactose and aspartame? Give an outline of how you would set this up as a practical investigation in the laboratory.

B3 Food storage

B3.1 Start your review of this topic at the end of the story . . . or *nearly* at the end.

Imagine you have some produce inside a package the material of which is *impermeable*, but transparent so you can see what is going on inside. For your produce you could choose a fruit or vegetable, such as apples, strawberries, tomatoes or lettuce.

(a)

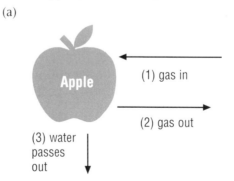

Refer to the numbered arrows.

(1) Which gas passes *into* the fruit? What happens when this gas drops to a low level in the package or is used up?

Unit 4

(2)   Which gas passes *out* of the fruit? What happens as this gas builds up inside the package?

(3)   How does the water passing out alter the relative humidity inside the package? How is this likely to affect the onset of microbial attack?

(b)   Suppose there is *no* packaging around the fruit. How would items (1), (2) and (3) affect the quality of the fruit and its storage time?

(c)   Now design suitable packaging for this fruit which would prolong its 'storage' or 'shelf' life and to help maintain its eating quality.

B3.2 Now go back to the beginning of the story.

(a)   You already know that **respiration** continues after harvest.

What effect does this respiration have on reserves of respiratory substrate within the produce?

How does carbon dioxide concentration (around the produce) affect postharvest changes? How does oxygen concentration (around the produce) affect postharvest changes?

How is the gas balance (of carbon dioxide and oxygen) important in controlled atmosphere (CA) and modified atmosphere (MA) storage? How does CA differ from MA?

How can packaging be designed to help maintain a desirable gas composition?

What effect would 'waxing' have on the internal atmosphere in the produce?

(b)   How does **loss of water** after harvest affect the produce? What steps can be taken to minimise water loss on storage?

(c)   Name *three* **sugars** commonly found in fruits. Name *two* **acids** commonly found in fruits as they ripen. Generally, how does the acid-sugar balance change as the fruit ripens? What effect does this balance have on taste of the produce? What other substances contribute to the flavours of the fruit?

(d)   **Ethene (ethylene)** is a plant growth substance. How does ethene affect the ripening process? In storage chambers, how can levels of ethene be controlled over a period of time and why is this necessary? What effect does a ripe banana have amongst green tomatoes (and explain your answer)?

(e)   What **colour changes** take place as a fruit ripens? Name the pigments linked to these colour changes.

B4 Biotechnology and food production

B4.1 Make yourself a summary of the following fermentations: *sauerkraut*; *soya sauce*; *wine*; *yoghurt*.
Make sure you know the **metabolic reactions** taking place during the fermentations – at least those which are significant in relation to the desired end-product – and collect together details about the **stages in the processing**.

Helpful hints

This series of questions takes you through some of the metabolic and other changes that take place in fruits and vegetables after harvest (postharvest changes). As you work through the questions, keep thinking about the effect on storage and on quality of the produce. Remember, some produce improves (for at least a period) after harvest, whereas other products start to deteriorate very soon.

In these questions, you can use apples or tomatoes as your examples.

Helpful hints

In B4.3 you may find it useful to organise your answers *either* in a table *or* you could devise a series of flow charts. Include information about temperature and the time for which the fermentation continues as well as the physical or other processing events.

You should be able to pick out the essential features of each fermentation and at the same time compare these across the different fermentations and their products. It is often useful to make these comparisons when learning. When you devise your table, or flow charts, make sure you arrange the items systematically and keep equivalent features together across the different examples.

Special hints for revision

This option has a wide range of topics so it is very important to learn the basic principles thoroughly. Many students' answers to questions are too vague: make sure that you know precise facts, such as definitions of BMI, lean body mass and the temperature at which pasteurisation is carried out. Some of the practical work could involve the use of complex or unfamiliar apparatus, so make sure that you know exactly how it works. For example, even if you have not used a calorimeter, find a diagram of one and an account of how it is set up so that you know the details. Be prepared for questions which present you with unfamiliar data or examples. There will always be enough information given to you in the question to enable you to apply your biological knowledge and write a sensible answer, provided that you have a sound understanding of the relevant topic.

Assessment questions

Mark allocations are given for each part of the questions and the answers are given on pages 125.

1   (a) Distinguish between the use of the resazurin test and the use of the turbidity test in relation to the freshness and storage of milk. **[3]**

    (b) Distinguish between the use of freezing and the use of canning in the long term storage of food. **[3]**

(Total 6 marks)
(Edexcel 6044/04, B/HB4D, January 2001, Q.2)

2   The photograph below shows a young child suffering from a metabolic disorder due to undernutrition.

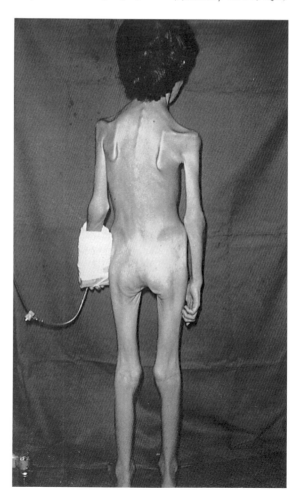

(a) State **two** symptoms of the disorder shown in the photograph. [2]

(b) Explain the causes of this types of metabolic disorder. [3]

(Total 5 marks)
(Edexcel 6044/04, B/HB4D, January 2001, Q.3)

3   Type II diabetes (mature onset diabetes) can usually be controlled by careful regulation of the diet. The table shows the nutritional content of some foods.

Food	Energy / kJ per 100 g	Protein / g per 100 g	Saturated fat / g per 100 g	Unsaturated fat / g per 100 g	Sugar / g per 100 g	Starch / g per 100 g
Rice	532	2.6	0.0	0.2	0.0	28.6
Breakfast cereal	1350	9.0	0.3	1.2	17.0	49.0
White bread	966	9.5	1.0	1.6	2.4	45.5
Young new potatoes	314	1.7	0.0	0.3	1.3	14.8
Muesli with dried fruits	434	10.7	5.8	13.9	20.0	37.9
Steamed pudding	1589	4.2	7.0	11.1	39.6	10.3
Pasta spirals	1538	11.5	0.0	1.5	0.0	75.7
Rolled oats	1564	10.0	1.9	5.5	0.1	71.1

(a) Suggest **two** of the foods in the table that would be most suitable for a person with type II diabetes to include as a major part of their diet. [2]

(b) Explain the reasons for your choice of these foods. [3]

(Total 5 marks)
(Edexcel 6044/04, B/HB4D, January 2001, Q.5)

4   The flow diagram shows some of the main stages in the production of soy sauce.

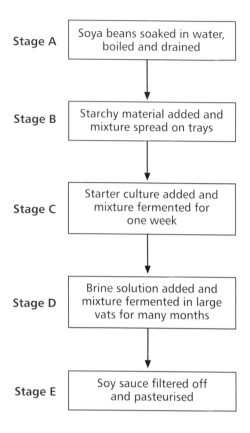

Stage A Soya beans soaked in water, boiled and drained

Stage B Starchy material added and mixture spread on trays

Stage C Starter culture added and mixture fermented for one week

Stage D Brine solution added and mixture fermented in large vats for many months

Stage E Soy sauce filtered off and pasteurised

(a) Suggest a suitable starchy material that could be added at stage B. [1]

(b) State the active components of the starter culture added at stage C. [1]

(c) Explain what happens to the starch during the fermentation at stage C. [2]

(d) (i) describe the conditions that are required for the fermentation at stage D. [1]

(ii) Name **one** chemical product of this fermentation that will contribute towards the preservation and the characteristic flavour of the soy sauce. [1]

(Total 6 marks)
(Edexcel, 6044/04, B/HB4D, January 2001, Q.4)

Human health and fitness

 Introduction

This option is divided into three sections:

C1 Body systems
C2 Exercise physiology
C3 Human disorders

In the first section, some of the topics in Units 2 and 4 are studied in greater depth in order to understand the effects of exercise, which are dealt with in the second section. There are also links between an understanding of the body systems and some of the disorders included in the third section.

Before revising this option in detail, it would be useful to review your understanding of related topics in the core units. These include the organisation of eukaryotic cells and transport across membranes (Unit 1.3), gas exchange in humans (Units 2B.1 and 2H.1), transport in mammals and blood and body fluids (Units 2B.2 and 2H.2), cellular respiration (Unit 4.1) and nervous coordination in mammals (Unit 4.2). Much of the material included in this option relies on a sound knowledge of these core topics and it is particularly important to remind yourself of the topics from the AS specification.

The checklist has been subdivided according to the sections of the option. It would be sensible to organise your revision so that you attempt one section at a time, bearing in mind that you will need to look back at a number of core topics.

Checklist of things to know and understand

Before attempting any of the questions or activities, check that you know and understand the following:

C1 Body systems

☐ the histology of cardiac muscle

☐ the structure of the mammalian heart and myogenic stimulation

☐ how the cardiac cycle is controlled, the roles of respiratory gases, control centres in the medulla, sympathetic and parasympathetic nerves

☐ the roles of the carotid sinus, aortic and Bainbridge reflexes

☐ the use of artificial pacemakers in the treatment of heart disease

☐ the roles of leucocytes in phagocytosis and secretion of antibodies

☐ the roles of lymphocytes in the immune response; active and passive immunity

☐ the structure of the breathing system and the mechanism of ventilation

☐ the histology of lung tissue

☐ how the ventilation mechanism is controlled; the roles of respiratory gases; control centres in the medulla, stretch receptors and cranial nerves

❑ the structure of compact bone and a synovial joint

❑ the structure and histology of striated muscle

❑ the structure and role of fast and slow muscle fibres

❑ the role of actin, myosin, calcium ions and ATP in muscle contraction

❑ the structure and role of the neuromuscular junction

❑ the structure of the lymphatic system, vessels, glands and connections with the cardiovascular system

❑ the formation and content of lymph

❑ the structure and role of lymph nodes in the immune response.

C2 Exercise physiology

❑ the roles of erythrocytes, haemoglobin and myoglobin

❑ the effect of skeletal muscle contraction on venous blood flow

❑ know that cardiac output is a function of heart rate and stroke volume

❑ the effect of exercise on cardiac output

❑ the meaning of the terms vital capacity and tidal volume

❑ the effect of exercise on breathing rate, tidal volume and residual volume

❑ know that minute volume (V_E) is a function of the breathing rate and tidal volume

❑ that ventilation uses oxygen and glucose

❑ the effect of training on ventilation efficiency

❑ the processes of aerobic and anaerobic respiration

❑ the role of muscle spindles in muscle contraction

❑ the speed force and fatigue characteristics of motor units

❑ the relationship between muscle structure and strength and the response of muscle to training

❑ the role of lactic acid in the production and elimination of an oxygen debt

❑ the nutritional requirements of a training programme

❑ the principles of aerobic training and its effect on cardiac output and oxygen transport

❑ the principles and effects of anaerobic conditioning and the role of creatine phosphate.

C3 Human disorders

❑ the causes and treatment of the cardiovascular disorders of atherosclerosis, hypertension and coronary heart disease

❑ the causes and treatment of the pulmonary disorders of bronchitis, TB, pneumoconiosis and lung cancer

❑ the causes and treatment of arthritis and osteoporosis.

Practicals

You are expected to have carried out the following practical work:

❑ *the study of prepared slides of cardiac tissue, lung tissue and striated muscle tissue*

❑ *the effect of ATP on the contraction of muscle fibres*

❑ *an investigation of the effects of physical activity on pulse rate and blood pressure*

❑ *the use of simple apparatus to estimate vital capacity and variation of breathing with physical activity*

❑ *an investigation of the effect of a training programme*

❑ *an estimation of percentage body fat using skinfold calipers.*

 Practical work – Helpful hints

Practical work should familiarise you with the structure of cardiac muscle, lung tissue and striated muscle tissue. You should also revise the relationship between structure and function of these tissues. The effect of ATP on contraction of muscle tissue is also investigated.

● Examine prepared slides of cardiac muscle, lung tissue, and striated muscle.

● Make annotated drawings of these tissues to explain the relationship between structure and function.

● Remember that a scale should always be included on a drawing to indicate the actual size of the specimen.

● To investigate the effect of ATP on muscle contraction, small strands of fresh muscle tissue are placed in warm Ringer's solution on a microscope slide.

● The length of the muscle strand is measured and recorded, then one drop of ATP solution is added.

● The length of the muscle strand is measured again.

You are also expected to investigate the effect of physical activity on pulse rate and blood pressure.

● For meaningful results, it is important to use some form of standardised activity, such as using an exercise cycle, or step-ups at a fixed rate.

● Pulse rate can be determined by placing the fingers over the radial artery in the wrist, or by using a digital pulse monitor.

● Blood pressure is measured using a sphygmomanometer.

● Blood pressure readings give two values, the systolic value and the diastolic value. Blood pressure is usually quoted in mm of mercury (mm Hg), although this is not an SI unit. To convert mm Hg to kilopascals (kPa), divide by 7.5.

● Record your resting pulse rate and blood pressure, then immediately after exercise.

You should be familiar with the use of simple apparatus to estimate vital capacity and variation of breathing with physical activity.

● This practical involves measurement of vital capacity and variation in breathing with physical activity.

● Vital capacity is defined as the maximal volume of air which can be expelled from the lungs by forceful effort following a maximal inspiration.

● Vital capacity can be measured using a spirometer, lung volume bags, or by displacement of water from a calibrated bell jar.

● Breathing rate can be determined by recording the number of breaths per minute.

- It is important to standardise the activity if meaningful results are to obtained.
- You could record changes in breathing rate at rest, then after cycling on an exercise bicycle at increasing speeds, such as 5, 10, 15, 20 and 25 km h^{-1}, for two minutes at each speed.

The investigation of a training programme gives you the opportunity to make some quantitative records of the effects of a training programme. This is a relatively long-term investigation and should be spread over at least two weeks. You could devise a specific training programme for your preferred sport, or use a programme which involves jogging specific distances within recommended times.

- Carry out your preferred training programme.
- Record your resting breathing rate and pulse rate, then immediately after the activity and again after a rest of one minute.
- Work out your recovery rate for each training session. Recovery rate can be expressed as pulse rate immediately after the activity - pulse rate after a one minute rest.
- Record all your results in a suitable table and discuss how the training programme affected your cardiovascular system.

This final practical involves measurement of percentage body fat using skinfold calipers. These are used to measure the thickness of a fold of skin with its underlying layer of subcutaneous fat. It should be noted, however, that this method may give unreliable results and more accurate data relating to body composition are obtained using the technique of bioelectric impedance analysis (BIA). This method involves passing a small electric current through the body and the level of body fat is calculated from the impedance to the flow of the electric current.

- Skinfold measurements provide a method for estimating the percentage of body fat.
- Skinfold calipers have a spring which exerts a standard pressure and a scale to measure skinfold thickness in millimetres.
- Measurements of skinfold thickness are taken at four sites, at the front of the upper arm, the back of the upper arm, on the back just below the shoulder blade, and on the side of the waist.
- These readings are added together, then Tables are used to convert to a percentage body fat.
- A less accurate estimation can be obtained using skinfold measurements from the back of the upper arm only. Again, these measurements are converted to percentage body fat using Tables.

Helpful hints

It might be useful to draw a table summarising the similarities and differences between cardiac, striated and smooth muscle

It might be useful to repeat this task to summarise the mechanism of expiration

C1 Body systems

1C.1 State *two* differences between cardiac muscle and striated muscle.

1C.2 Construct a flow diagram to summarise the mechanism of inspiration.

1C.3 Copy the description of a neuromuscular junction given below and then write on the dotted lines the most appropriate word or words to complete the description.

"A neuromuscular junction is the point at which a motor .... connects with a muscle. Impulses are transmitted to the muscle by .... (a neurotransmitter). This triggers an .... in the muscle, leading to ...."

C2 Exercise physiology

2C.1 Cardiac output (Q) is related to stroke volume (SV) and heart rate (HR) according to the formula: $Q = SV \times HR$. Calculate the heart rate when $Q = 5040$ cm^3 min^{-1} and $SV = 70$ cm^3.

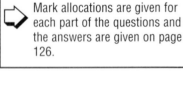

Mark allocations are given for each part of the questions and the answers are given on page 126.

Helpful hints

This section of the specification requires you to know the causes of each disorder and the relevant *treatment*. A table summarising the causes and treatments of each disorder would be useful for revision.

2C.2 What are the two possible fates of the lactic acid produced by anaerobic respiration?

2C.3 Explain what is meant by the term *glycogen loading*.

C3 Human disorders

3C.1 Identify *three* risk factors associated with the development of atherosclerosis.

3C.2 What is pneumoconiosis?

3C.3 Outline the methods used to treat lung cancer.

Special hints for revision

The 'Body systems' section of this option is extensive and it would be helpful to revise each system separately at first, making reference to the relevant topics in the core units. Many of the topics listed in the checklist will be familiar to you and are really reminders that you should know these areas thoroughly. For example, you will already have learnt about the cardiac cycle, the functions of leucocytes and the mechanism of ventilation, but you need this knowledge as a starting point for the special option topics. There are important links between the cardiovascular and pulmonary systems, especially with respect to the roles of respiratory gases and the nervous system in their control. There are also links between the cardiovascular system and the lymphatic system. You should be aware of these inter-connections and how they are relevant to the section on exercise physiology. You could try devising revision activities which make you integrate your knowledge so that you appreciate that exercise affects more than one body system at a time. The practical work is closely related to the specification topics and should help you with your understanding. It is a good idea to refer to any practical work that you have done particularly for the exercise sections, as many data interpretation questions could well be similar to experiments that you have carried out.

Assessment questions

1   The graph below shows the relationship between the length of a sarcomere and the tension developed in a myofibril in striated muscle.

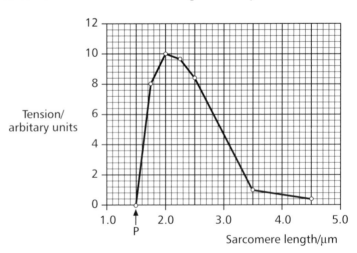

(a) From the graph, find the sarcomere length at which the maximum tension is developed. [1]

(b) The diagram below shows the structure of a part of a myofibril. Name the parts labelled A, B and C. [3]

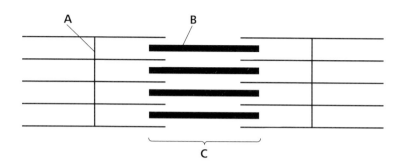

(c) In the space below, make a drawing to show the same part of the myofibril as it would appear at position P indicated on the graph. [2]

(Total 6 marks)
(Edexcel 6050, HB3, June 2001, Q.5)

2 Describe the causes of each of the following.
   (a) Bronchitis. [2]
   (b) Tuberculosis (TB). [2]
   (c) Pneumoconiosis. [2]

(Total 6 marks)
(New question)

3 (a) Distinguish between active and passive immunity. [3]
   (b) Describe the role of T cells in the immune response. [3]

(Total 6 marks)
(New question)

4 An investigation was carried out on the effect of exercise on heart rate and stroke volume. The heart rate is the number of heart beats per minute and the stroke volume is the volume of blood pumped out by the heart each time it beats. The heart rate and stroke volume of a person were measured at different work levels. The results of the investigation are shown in the graph below.

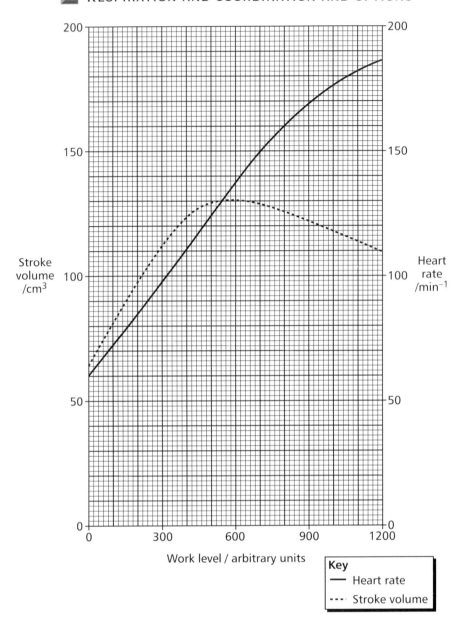

(a) Describe and comment on the changes in stroke volume with increasing levels of work. **(3 marks)**

(b) Calculate the percentage increase in heart rate as the work level increases from 300 to 900 arbitrary units. Show your working. **(3 marks)**

(c) Cardiac output is the volume of blood pumped out by the heart in one minute. Cardiac output can be calculated using the following formula.

Cardiac output = stroke volume × heart rate

Describe how the cardiac output changes as the work level increases from 0 to 1200 arbitrary units. **(2 units)**

(d) Describe the mechanisms which bring about an increase in heart rate with increasing work level. **(4 marks)**

(Total 12 marks)

(Edexcel 6050, HB3, June 1997, Q.7)

5B Genetics, evolution and biodiversity

Introduction

Unit 5B is the Biology pathway. If you are studying Biology (Human) go to page 77.

This unit contains a number of different topics which bring together many aspects of Biology and there is development and extension of concepts and principles studied in other units of the specification. The first two topics cover the important metabolic pathway of photosynthesis and the way in which control of growth in plants is achieved. Some understanding of classification is required as an introduction to the distribution of plants and animals, populations and succession. The rest of the unit is mainly about the transmission of genetic information from generation to generation and how this process maintains the continuity of species. It also examines how genetic differences can lead to natural selection and the evolution of the great diversity of living organisms that we see throughout the world. In order to understand this biodiversity, it is also important to consider the interaction between genes and the environment, so there is a link within the unit to the aspects of ecology which are addressed. Finally, the manipulation of DNA, the introduction of new genes into organisms and some applications of gene technology are introduced.

At all stages of your study and learning of the material contained within this unit, you should be prepared to review topics from the AS units. This is particularly important as the written test will include synoptic questions, requiring you to use knowledge gained from the AS units in your answers. The introduction to each topic outlines what you should know from studying previous units.

The unit is divided into four topics:

1B Photosynthesis
2B Control of growth in plants
3B Biodiversity
4B Genetics and evolution

Topic **1B** Photosynthesis

Introduction

The process of photosynthesis is an important metabolic pathway and involves a number of concepts and principles already covered in other units of the specification. Before studying the process in depth, it is advisable to have a sound understanding of the nature of biological molecules, such as carbohydrates, lipids and proteins, the structure of leaves and a knowledge of the transport mechanisms in flowering plants. As much of this is covered in the first three units of the specification, the relevant topics and the unit in which they appear are summarised below.

- **From Unit 1:**
  - carbohydrates, lipids and proteins
  - enzymes
  - the structure of a leaf palisade cell
  - chloroplast structure
  - the distribution of tissues in a mesophytic leaf.

- **From Unit 2:**
  - gas exchange in flowering plants
  - structure and roles of stomata
  - roles of xylem and phloem in transport
  - uptake and transport of water;
  - transport of mineral ions through the plant.

- **From Unit 3:**
  - the role of producers in ecosystems.

- **From Unit 4:**
  - the role of producers in ecosystems.

In Unit 5B, you are required to study the main chemical reactions involved in the light-dependent and the light-independent reactions. The level of knowledge needed is clearly described in the specification and you will only be asked about those chemical reactions which are stated. Details of intermediate compounds and individual reactions will not be expected.

When studying this topic, there is plenty of opportunity for practical work, and carrying this out and learning the details of the experiments helps you to understand the stages of photosynthesis and the factors which affect it. You could be asked to describe the apparatus used, give practical details, such as precautions and conditions, and interpret experimental data. So, even if you are not able to carry out all the required practical work personally, you should make sure that you know how the apparatus works and all the details.

 Checklist of things to know and understand

Before attempting to answer any of the questions, check that you know and understand the following:

Photosynthesis

❏ that photosynthesis is the synthesis of organic compounds as a result of the fixation and reduction of carbon dioxide

Leaf structure

❏ the external and internal structure of a dicotyledonous leaf

❏ the location of the palisade tissue

❏ the structure of a palisade cell

❏ the structure of a chloroplast as revealed by electron microscopy

❏ the nature and location of the chloroplast pigments

❏ the absorption and action spectra for chloroplast pigments

Light-dependent reaction

❏ the processes of cyclic and non-cyclic photophosphorylation and the production of reduced NADP, ATP and the evolution of oxygen related to the light-independent reaction

❏ the evolution of oxygen

Light-independent reaction

❏ the fixation of carbon dioxide on to a 5-C compound (ribulose bisphosphate) to give phosphoglyceric acid (PGA)

❏ the use of reduced NADP and ATP in the synthesis of carbohydrate from PGA

❏ the regeneration of the 5-C compound

Environmental factors affecting the rate of photosynthesis

❏ the effect of light intensity, wavelength, carbon dioxide concentration and temperature on the rate of photosynthesis

❏ the concept of limiting factors

❏ the compensation point

Mineral nutrition

❏ the functions of nitrate, phosphate and magnesium ions

❏ uptake of mineral ions by roots.

Practicals

You are expected to have carried out practical work to investigate:

❏ *the chromatography of chloroplast pigments*

❏ *the effects of light intensity and carbon dioxide concentration on the rate of photosynthesis*

❏ *the effect of mineral ions on plant growth using mineral culture solutions.*

〖〗〗 *Practical work – Helpful hints*

In this section, you should be familiar with the extraction of chloroplast pigments and their separation using the technique of chromatography. The solvents used are volatile and highly flammable, so it is advisable to carry out the practical in a fume cupboard.

- Chloroplasts contain a mixture of pigments including chlorophyll a, chlorophyll b, xanthophylls and β-carotene.
- Pigments are extracted by grinding leaf tissue with a pestle and mortar in a solvent such as propanone.
- The extracted pigments are then applied to chromatography paper or a thin-layer chromatography plate, using a fine pipette. The application is repeated to produce a small, concentrated spot of pigments.
- An organic solvent mixture, such as hexane-ethoxyethane-propanone, is then used to separate the pigments. As the solvent mixture rises up the chromatography paper or the thin-layer plate, the different pigments move at different speeds and so separate out.
- The different pigments can be identified by their colour and Rf values.
- The Rf value is the distance moved by the pigment spot divided by the distance moved by the solvent.

You are also expected to have made quantitative measurements of the effects of light intensity and carbon dioxide concentration on the rate of photosynthesis. An aquatic plant, such as *Elodea*, is used for this investigation as the oxygen produced in photosynthesis is given off as bubbles of oxygen gas. The bubbles can be either counted or collected and the volume measured.

- The number of bubbles produced per unit time, or the volume of oxygen collected per unit time, gives a measure of the *rate* of photosynthesis.
- The rate of photosynthesis is influenced by a number of different factors, so careful experimental design is important so that only one variable changes. Other factors must be kept constant.
- To vary light intensity, the light source can be moved different distances away from the plant. Light intensity is inversely proportional to the square of the distance (the inverse square law).
- To vary carbon dioxide concentration, you can use solutions of sodium hydrogencarbonate ($NaHCO_3$) of known concentrations, such as 0.1, 0.2, 0.3, 0.4 and 0.5 %.

You should also have carried out an investigation into the effects of mineral ion deficiency on the growth of plants. Mineral solutions containing a range of mineral salts are used and the effects on the growth of seedlings are noted.

- Seedlings are grown in tubes containing a range of mineral salts, including those lacking specific ions, such as phosphate, nitrate, sulphate and magnesium.
- The tubes are covered with aluminium foil or black paper to prevent the growth of algae in the mineral solutions.
- Suitable seedlings to use in this investigation include maize, castor beans, tomato and cabbage.
- The seeds are first germinated in moist vermiculite, then the seedlings are transferred separately to the mineral solutions.
- At weekly intervals, the plants are measured and any deficiency symptoms noted.

Unit 5B

 Testing your knowledge and understanding

The answers to the numbered questions are on pages 111.

To test your knowledge and understanding of leaf structure and the involvement of the structures in the leaf in the process of photosynthesis, try answering the following questions.

1B.1 The diagram below is a transverse section through a dicotyledonous leaf.

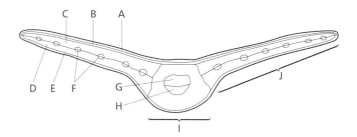

(a) Name the structures labelled A to J.

(b) Identify the following structures from the description of their function.

- **(i)** The site of most photosynthetic activity
- **(ii)** The region where most gaseous exchange takes place
- **(iii)** Transparent layer allowing light through to the mesophyll
- **(iv)** Conducts water and mineral ions to the leaf
- **(v)** Broad, thin and flat, providing a large surface area for light absorption
- **(vi)** Supports the leaf tissues
- **(vii)** Transports the soluble products of photosynthesis away from the leaf
- **(viii)** Reduces water loss from the leaf tissues

1B.2 Make a drawing from memory of the structure of a chloroplast as revealed by electron microscopy. On your diagram, label the chloroplast envelope, the stroma, intergranal lamellae, grana / thylakoids, starch grains, lipid droplets.

Annotate each of the labels on your diagram with the function of the structure or region.

1B.3 The following equation summarises the process of photosynthesis.

$$CO_2 + 2H_2O \longrightarrow (CH_2O) + O_2 + H_2O$$

(a) State **two** conditions that are necessary for this process.

(b) Where does the carbon dioxide come from and how does it get to the photosynthesising cells?

(c) During which part of the process are the water and the oxygen formed?

(d) Name the process by which the carbon dioxide is taken up by the 5-C compound.

(e) Name the first product of photosynthesis from which carbohydrates are formed.

Helpful hints

When you have tried out these questions as they are printed here, it would be a good idea to review them at a later date, but this time see if you can decide what the questions were from the answers at the back of the book. It would be possible to test your knowledge of the following.

---

1 Chlorophyll pigments, absorption spectra and action spectra

● List all the pigments, describing their colours and the wavelengths of light which they absorb.

● Draw an action spectrum and an absorption spectrum for the pigments. Make sure you know the difference between the two.

● Find a diagram which shows the location of the pigment molecules on the lamellae. It is unlikely that you would be asked a detailed question about the arrangement of the pigment molecules but it helps you to visualise what is going on inside a chloroplast.

---

2 Light-dependent reaction / cyclic and non-cyclic photophosphorylation

● Cyclic photophosphorylation can be represented as a cycle of events. Try drawing out this cycle and putting as much information on it as possible, such as which photosystem is involved and what is produced.

● Write an equation summarising the events of non-cyclic photophosphorylation.

● Make a flow chart to show the relationship between the two photosystems and the pathways of the electrons emitted from the chlorophyll molecules.

---

3 Light-independent reaction

● Try building up a cycle of events beginning with the 5-C compound and carbon dioxide. Show on your diagram the enzyme responsible for carboxylation, where ATP is used, where reduced NADP is needed, the formation of complex carbohydrates and the regeneration of the 5-C compound.

---

4 Environmental factors

Sketch a graph of the way in which the rate of photosynthesis varies with increasing light intensity and then consider the effects of increasing the temperature and the carbon dioxide concentration. Additional curves can be added to your original graph. Try giving a full explanation for each of the curves you put on the graph. Some of these activities are suitable for revision cards as they summarise the events of the process and factors which affect it.

 Mark allocations are given for each part of the questions and the answers are given on pages 111–12.

Helpful hints

Make sure that the feature you choose and your explanation are full and relevant.

You could extend this question by labelling everything you can see on this photomicrograph and by finding as many ways as possible in which the cells are adapted: there are at least five.

 Practice questions

The questions given here are related to the content of Unit 5. Other questions, involving different skills and including reference to other topics in different units, may form part of the synoptic assessment (see page 00).

1 The photomicrograph below shows the structure of part of a leaf as seen in transverse section.

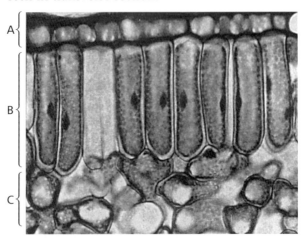

Unit 5B

(a) Name the parts labelled A, B and C. **[3]**

(b) Explain **two** ways in which the cells in part B are adapted for their function. **[4]**

(Total 7 marks)
(Adapted from Edexcel 6042, B2, June 1997, Q.1)

2   The diagram below shows some of the processes which occur in the light-independent reaction of photosynthesis.

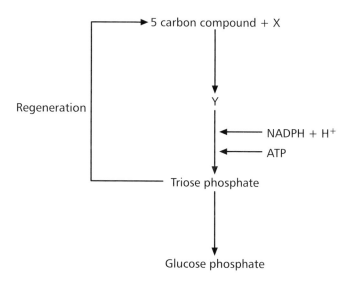

(a) Name the substances represented by the letters X and Y. **[2]**

(b) State the origin of the NADPH + H⁺ and the ATP used in the light-independent reaction. **[1]**

(c) Where in the chloroplast does the light-independent reaction occur? **[1]**

(Total 4 marks)
(Edexcel 6042, B2, January 1997, Q.4)

3   The statements in the table below refer to the light-dependent and light-independent reactions of photosynthesis.

Helpful hints

Remember to make your ticks and crosses very clear. If you change your mind cross out your first answer and write your amended answer by the side. Don't try to change a cross into a tick or vice versa.

Statement	Light-dependent reaction	Light-independent reaction
Oxygen produced		
Carbon dioxide fixed		
Occurs in stroma		
Uses NADPH + H⁺		
Produces ATP		

If the statement is correct for the process, place a tick (✓) in the appropriate box and if it is incorrect, place a cross (✗) in the appropriate box.

(Total 5 marks)
(Edexcel 6042, B2, June 1996, Q.4)

4   The apparatus shown below can be used to study mineral nutrition in flowering plants.

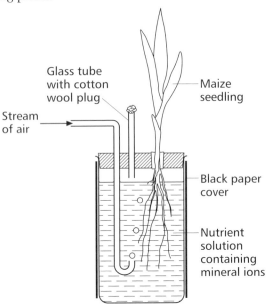

(a) Suggest the function of each of the following.

(i)   The stream of air   **[2]**

(ii) The black paper cover   **[2]**

(b) The nutrient solution contains various mineral ions including magnesium and phosphate.

Give **one** reason why each of these ions is essential to the plant.   **[2]**

(c) Describe how you could use such apparatus to investigate the effects on the growth of maize seedlings of different concentrations of nitrate ions in the nutrient solution.   **[4]**

(Total 10 marks)

(Adapted from Edexcel 6042, B2, January 1996, Q.5)

5   An investigation was carried out into the effect of light intensity on the uptake and release of carbon dioxide by two green plant species J and K.

Single leaves of each species, still attached to plants, were sealed in glass vessels containing a known mass of carbon dioxide. The leaves in the vessels were exposed to light of known intensity for one hour. The change in mass of carbon dioxide in each vessel was determined and the change in mass of carbon dioxide per cm^2 of leaf surface was then calculated. The experiment was repeated at a range of light intensities for both species.

The results are shown in the graph opposite. Light intensity is expressed as a percentage of normal daylight.

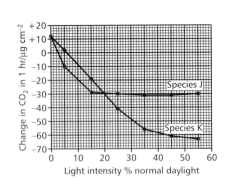

(a) (i)   From the graph determine the light intensity at which there is no net exchange of carbon dioxide by the leaves of species J and K.   **[2]**

(ii) Describe and explain the relationship between light intensity and the exchange of carbon dioxide in species J.   **[3]**

(b) (i)   Give **two** differences between the curves for species J and K.   **[2]**

> **(ii)** Suggest which species would be better adapted to a shaded habitat giving a reason for your answer. **[2]**
>
> **(c) (i)** State **one** environmental factor, other than light intensity, which could affect the uptake and release of carbon dioxide in the two plant species. **[1]**
>
> **(ii)** Explain why the factor you have chosen in (i) has an effect on carbon dioxide uptake and release. **[2]**
>
> **(Total 12 marks)**
> *(Edexcel 6042, B2, January 1996, Q.6)*

Topic 2B

Control of growth in plants

 Introduction

This topic recalls the detection of light by flowering plants studied in Unit 4 and looks at the effect of light on the growth of plants. In addition, you are expected to be able to explain the ways in which plant growth is regulated by five groups of plant growth substances: auxins, cytokinins, gibberellins, abscisic acid and ethene. You need to have studied that part of Topic 2 in Unit 4 that deals with response to changes in the external environment and it would also be helpful to know about photosynthesis (Topic 1B in Unit 5B) It would be relevant to try to work out the similarities and differences between these plant growth substances and the nature and action of animal hormones. An understanding of the commercial applications of auxins is also required.

> You should link your revision in this topic to aspects of Unit 4 (Response to changes in the external environment) and Unit 5B, Topic 1B (Photosynthesis).

 Checklist of things to know and understand

Before attempting to answer any of the questions, check that you know and understand the following:

- ❏ that phytochrome pigments are involved in the detection of light in flowering plants

- ❏ how light affects the growth of plants

- ❏ the nature of plant growth substances and the specific effects of auxins, cytokinins, gibberellins, abscisic acid and ethene

- ❏ the ways that auxins have been exploited through different commercial applications

- ❏ what is meant by synergism and antagonism in the context of plant growth substances.

Practicals

You are expected to have carried out practical work to investigate:

- ❏ *the effect of plant growth substances, such as rooting powder and weedkillers, on plant growth.*

Practical work – Helpful hints

The aims of this practical are to investigate some of the commercial applications of plant growth substances. Some synthetic auxins are used to encourage rooting in cuttings, and others are used as selective herbicides, for example to kill broad-leaved weeds growing in a cereal crop.

- Rooting powders usually contain the synthetic auxins indole butyric acid (IBA) or α naphthalene acetic acid (NAA).
- Flower pots are filled with a suitable compost and cuttings dipped in rooting powder.
- Suitable cuttings for this investigation include *Coleus, Erica, Fuchsia*, mint or willow.
- The treated cuttings are inserted into the compost and covered with a polythene bag.
- A control should be included, with untreated cuttings.
- Synthetic auxins used as selective herbicides include 2,4-D (2,4-dichloro-phenoxyacetic acid) and MCPA (4-chloro-2-methylphenoxyacetic acid).
- You can investigate the effect of these herbicides by spraying plants such as barley and radish, growing in compost in a plastic seed tray.
- Suitable hand sprayers containing selective herbicides are available from garden centres.
- Always follow the safety instructions carefully. Keep synthetic auxins off your skin, wash off any accidental spillage. Do not breathe the spray and wash hands and exposed skin thoroughly after use.

 Testing your knowledge and understanding

The answers to the numbered questions are on pages 112–13.

To test your knowledge and understanding of the control of growth in plants, try answering the following questions.

2B.1 Light can affect, regulate or control the growth of plants in different ways. Explain how each of the following can affect plant growth: **intensity** of light, **direction** of light, **duration** of light (**daylength**). In each case, link the effect with one of the following terms: *photoperiodism; photosynthesis; phototropism*.

2B.2 The numbered statements (1 to 32) below refer to different plant growth substances [**auxins, cytokinins, gibberellins** (e.g. **GA₃**) **abscisic acid** (or **ABA**) and **ethene**]. Make a table and list the plant growth substances in one column, then match the statement or description to the correct plant growth substance in the other column. Some descriptions fit more than one of the plant growth substances.

Helpful hints

When you have completed the table, you will have a useful summary of the information relating to different aspects of plant growth substances.

List of statements relating to features of plant growth substances:

1   is a gas

2   one example, zeatin, is found in endosperm of maize and the liquid endosperm of coconuts

3   associated with abscission of leaves at leaf fall and of ripe fruits

4   first discovered from the effects of a fungus infection of rice plants

5   activity linked with tropic responses

6   activity often linked with environmental stress

7   promote stem elongation in genetically dwarf plants

8   shows marked increase during ripening in certain fruits, often associated with rise in respiration

9   involved with dormancy in seeds and buds

10   can make long-day plants bolt in short days (overcoming photoperiodism effects)

11   promotes the activity of cellulases and polygalacturonase

12   are synthesised in regions of cell division and enlargement

13   stimulates cell division

14   found mainly in young meristems

15   inhibits growth of plants

16   promote growth (elongation of coleoptiles) and stems

17   delay senescence

18   promote growth and development of fruits (e.g. can increase the size of seedless grapes)

19   promote growth by increasing plasticity of cell wall (enabling it to take up more water and expand)

20   involved with closure of stomata in times of water deficit

21   break dormancy in seeds

22   promotes the activity of cellulases and polygalacturonase and is linked to events during the ripening of fruits

23   used in tissue culture media

24   curvature of shoots towards light results from a redistribution of this plant growth substance laterally across the stem

25   auxin stimulates its synthesis

26   activity often related to nucleic acid metabolism and protein sythesis

27   stimulate aleurone layer in germinating barley seeds to release α-amylase

28   inhibits growth which is promoted by auxin

29   promotes abscission but auxins prevent it

30   acts as inhibitor in transitions from dormancy to germination, whereas cytokinins and gibberellins act as growth promoters

31   accelerates senescence, whereas cytokinins delay senescence

32   effects often depend on other plant growth substances, such as auxin

2B.3 Look again at the list of descriptions given above.

    (a) Which of these plant growth substances *promote* growth and which usually inhibit growth?

    (b) Find examples of where two (or more) plant growth substances act *synergistically* and where others act *antagonistically*.

2B.4 The following practices are used commercially in horticulture and / or agriculture. How are these applications linked to activity of auxins (natural or synthetic)?

    (a) use of 'hormone' rooting powder

    (b) pruning

    (c) promoting development of fruits

    (d) keeping fruits on trees until harvest

    (e) weedkillers

Practice questions

1   Experiments were carried out to investigate the effect of light on the distribution of auxin (IAA) in coleoptile tips of maize seedlings. Four experiments (A, B, C and D) were set up in which the coleoptile tips were treated as described below.

The results are shown in the diagrams below. The relative auxin concentrations that collected in the agar blocks are shown as percentages.

Treatment	**Result**

Experiment A
Intact coleoptile tips were illuminated uniformly.

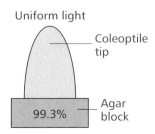

Experiment B
Intact coleoptile tips were kept in the dark

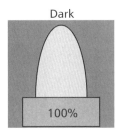

Experiment C
Coleoptile tips were partially divided by a piece of microscope coverslip glass, which also divided the agar blocks on which the tips stood. Tips were illuminated from one side.

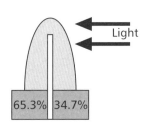

Experiment D
Coleoptile tips and agar blocks were completely divided by microscope coverslip glass. Tips were illuminated from one side.

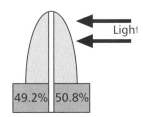

(a) Compare the results of Experiment C with those of Experiment D. Suggest an explanation for the differences. **(3 marks)**

(b) Do the results of these four experiments support the hypothesis that light promotes the destruction of auxin? Give an explanation of your answer. **(3 marks)**

(c) Describe **two** ways in which synthetic auxins are used commercially. **(4 marks)**

(Total 10 marks)

(Edexcel 6043/01, B3, June 2001, Q.6)

2 The table below refers to plant growth substances and their functions.

Complete the table by inserting the correct word or words in the blank spaces.

Plant growth substance	*One* function
Auxin	
	Breaks seed dormancy Causes bolting in LDPs
Ethene	
	Inhibition of leaf ageing
	Promotion of fruit fall

(Total 5 marks)
(Edexcel 6043, B3, June 2000)

Mark allocations are given for each part of the questions and the answers are given on pages 113.

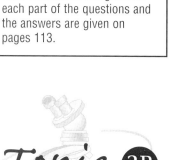

Topic 3B Biodiversity

 *Introduction*

Although this topic is entitled Biodiversity, it looks beyond the range of living organisms and how they are classified into taxonomic groups. You are required to have an understanding of the effects of abiotic and biotic factors on the distribution of organisms in terrestrial and aquatic habitats. This can be achieved by practical investigations that involve the use of qualitative and quantitative sampling techniques. It is important to relate your revision of this topic to your practical work and to a specific habitat, as you should be familiar with the organisms in the habitat you studied and with the techniques you used. It would be helpful to link you revision with a review of Topic 3 in Unit 2B, where adaptations of organisms to the environment were considered.

A more general review of Unit 3, particularly Topics 3.2 (Ecosystems), 3.3 (Energy flow) and 3.6 (Human influences on the environment) would be extremely useful before starting your learning of the rest of Topic 3B. A sound knowledge of the meanings of the terms used when discussing populations is important.

There is a substantial amount of practical work associated with this topic, which can help you in your understanding of the details you need in order to be able to answer questions successfully. You may be asked about habitats that you have not studied or be required to analyse unfamiliar data, so the more thorough your understanding, the better prepared you will be.

You should link your revision in this topic to Unit 3 in your AS course, in which you study ecosystems and energy flow and also to the section in Unit 2 on adaptations shown by organisms to their environment.

Checklist of things to know and understand

Before attempting to answer any of the questions, check that you know and understand the following:

Classification

❏ that species are classified into groups using shared derived features

❏ the principles and importance of taxonomy based on kingdom, phylum, class, order, family, genus and species

❏ the important features of the five kingdoms

Distribution of plants and animals

❏ the effects of biotic and abiotic factors on the distribution of organisms in a terrestrial and aquatic habitat

❏ appropriate qualitative and quantitative field techniques used in an investigation of the distribution of organisms in a specific habitat

Succession

❏ that ecosystems are dynamic and subject to change over time, illustrated by the change from grassland or abandoned farmland to woodland

❏ the seral stages in a succession

❏ plagio and climatic climax

Populations

❏ the terms population, community, population size and density

❏ the factors which affect population size, in relation to carrying capacity, environmental resistance

❏ density-dependent factors and density-independent factors

❏ intraspecific and interspecific competition

❏ the possible effects of predator-prey relationships on population size

❏ how insect populations can be controlled by biological and chemical methods

❏ the relative advantages and disadvantages of these methods

❏ the bioaccumulation of non-biodegradable toxins

❏ the use of integrated pest management (IPM)

Conservation

❏ the management of grassland and woodland habitats to maintain or increase biodiversity as illustrated by mowing, scrub clearance, use of fire and coppicing

❏ how intensive food production may affect wildlife

❏ how farming practice can enhance biodiversity

❏ the significance of the EU Habitats Directive concerning the conservation of natural habitats and of wild flora and fauna

❏ the significance of Natura 2000.

Practicals

You are expected to have carried out practical work to investigate:

❏ *the distribution of plants and animals in at least one habitat*

❏ *the influence of abiotic factors on them*

❏ *population size using the Lincoln Index.*

⚗ *Practical work – Helpful hints*

You are expected to use sampling methods to study the distribution of plants and animals in at least one habitat. The nature of this investigation will depend on the types of habitats you study.

● Plan your investigation carefully before you start.

● Be aware of safety issues – this is particularly important when working in aquatic habitats.

● Decide whether you are going to use random sampling or systematic sampling.

● Quadrat frames (for example, 0.25 m²) are used to sample the area under investigation.

● When describing vegetation, you may decide to use a subjective estimate of percentage cover or an objective estimate, such as percentage frequency or biomass.

● For sampling in aquatic habitats, techniques such as sweep sampling, or kick sampling are used.

● Terrestrial animals may be collected using a Tullgren funnel or a pitfall trap.

● Environmental measurements in terrestrial habitats include edaphic factors, such as soil depth, moisture content, humus content, pH and mineral content.

● Environmental factors in aquatic habitats include substrate type, temperature, current velocity, light, and dissolved oxygen.

You should also have carried out practical work to estimate population size using the Lincoln index. The capture–mark–recapture method is a technique for estimating the size of a population of organisms. The principle of this method is that a sample of the population is taken and these organisms are marked in some way so that they can be identified later. These organisms are then released and allowed to disperse throughout the population. A second sample is then taken and the numbers of marked organisms recaptured, and those captured that are unmarked, are recorded. A formula, known as the Lincoln Index, is then used to estimate the total population size.

● There are a number of important assumptions in using this method:

◆ the mark has no effect on the organisms

◆ the mark persists during the investigation

◆ the marked organisms disperse randomly throughout the whole population

◆ the population is closed, that is, there is no migration of animals

◆ there are no births or deaths during the investigation.

The formula for the Lincoln Index is shown below:

$$N = \frac{S_1 \times S_2}{R}$$ where N = the estimated total population size

S_1 = the number of organisms marked and released

S_2 = the number of organisms captured in the second sample

R = the number of marked organisms recaptured

Unit 5B

 The answers to the numbered questions are on pages 114–17.

Testing your knowledge and understanding

To test your knowledge and understanding of biodiversity, try answering the following questions.

Classification

3B.1 (a) This question gives you some organisms for you to organise into a classification and to think about what the different taxonomic groups mean.

List A gives seven **taxonomic terms** which represent different levels in a hierarchy by which living organisms are classified.

List A: *class ; family; genus ; kingdom ; order ; phylum ; species ;*

Make a table with five columns. In the first column (A), arrange these terms in their correct taxonomic sequence. Look next at lists B, C, D and E. These show the full taxonomic descriptions of four different organisms (nettle, brooding star, brewing yeast, humans), but the descriptive terms are given in alphabetical order rather than in a logical taxonomic sequence. Arrange these in the remaining four columns of the table (headed B, C, D and E) so that they match the taxonomic levels given in column A.

List B: Nettle	Angiospermophyta; Dicotyledones; *dioica*; Plantae; Urticaceae; *Urtica*; Urticales;
List C: Brooding star (sea star)	Animalia; *Asterina*; Asterinidae; Echinodermata; *phylactica*; Stelleroidea; Spinulosida;
List D: Yeast (brewing)	Ascomycetes; Ascomycota; *carlsbergensis*; Endomycetales; Fungi; *Saccharomyces*; Saccharomycetaceae;
List E: Humans	Animalia; Chordata; Hominidae (hominids); *Homo*; Mammalia; Primates; *sapiens*;

(b) In the twenty-first century, what is likely to become increasingly important as a means of classification, or diagnosis of taxonomic (genomic) similarities?

(c) Using the information in the table, answer these questions.

(i) What features do the brooding star (list C) and humans (list E) have in common to put them in the *same* kingdom? What features in nettles (list B) and yeast (list D) put them in *different* kingdoms?

(ii) Which kingdom(s) is (are) not represented in these lists? Identify the features which would be typical of each of these kingdoms. Name *one* organism in each of the kingdoms *not* included in these lists, and identify the features that would be typical of each of these kingdoms.

Distribution of plants and animals

3B.2 Think of any **abiotic** factors that could influence plant and animal life in a habitat.

(a) Make a list of these abiotic factors and then arrange your list under two headings, one for those which would be relevant in *terrestrial* habitats and the second for those in aquatic habitats. Say whether your aquatic habitat is freshwater or marine. Do some factors fit in both lists?

(b) From your list, choose *two* (or more) factors that would be relevant in a habitat you have studied or visited. For each of these factors, describe how you would measure changes in this factor.

(c) Next relate these same factors to a particular habitat (and remember to *name* or describe essential features of the habitat) and describe how the factor might change over a period of time.

(d) Then link this factor to particular (*named*) examples of organisms, plant and animal, and describe any adaptations which may help the organism to live in that habitat.

3B.3 You need to be able to describe the qualitative and quantitative **field techniques** you could use in a particular habitat.

There are certain principles which are common to most habitats, so you can draw up a revision scheme for yourself which is appropriate for the area you have studied.

These questions can be adapted to act as 'prompts' for your own revision checklist.

(a) What initial observations would you make when approaching an investigation in a new area or habitat?

(b) Describe an appropriate sampling technique you would use. Is it a random or systematic sampling method? What apparatus do you use to limit the 'sample'?

(c) How do you identify *what* organisms are there?

(d) How do you make quantitative estimates – generally, then more precisely – within your sampled area?

(e) Describe how you would do counts of certain animals at different times. (This could, for example, be times of day or different seasons.)

(f) How would you record then display your results?

(g) How could you make comparisons of two different areas, or the same area over a period of time?

(h) Give an indication of the sort of data you would collect to ensure that you could determine whether differences are statistically significant. Follow through the planning you would do to ensure that this approach is integrated into your investigation from the start.

Unit 5B

Helpful hints

It may be tempting just to say 'use a probe' for each factor, but try to think of different practical techniques you may have used. Remember to make sure you know the relevant *units* for any measurements that you would make of your chosen abiotic factors.

'Field techniques', to put it simply, means you need to record *what* is there (plants and animals and perhaps seaweeds, fungi or lichens), *how much* of it is there and *where* it is within the area you have chosen for study. The techniques you use depend very much on the area being investigated and the nature of the organisms there.

In (a) it may be a good plan to start with a terrestrial habitat (and this could be a marine littoral habitat – e.g. the sea shore), then look for equivalent techniques that could be used in an aquatic habitat.

Succession

3B.4 The diagrams 1 to 4 below represent four stages in a **succession** from grassland to woodland. From the diagrams, or from your own knowledge of a similar area, answer these questions.

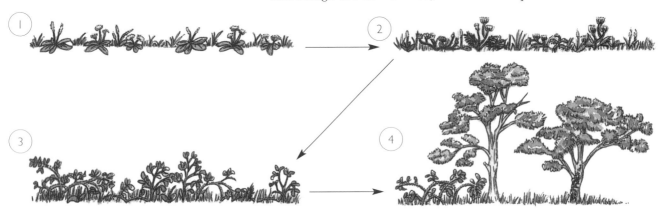

Helpful hints

Compare stage 1 with stage 2, then compare stage 3 with stage 2 and finally look at stage 4 compared with stage 3.

You would probably relate the time scale to a temperate climate but give another if that is appropriate to your studies.

Give examples of likely plant species in the different stages if you can.

(a) Describe how the plants in each stage differ from those in the stage before. Your answers may refer to habit of the plant as well as species diversity.

(b) What factors could prevent stage 1 developing into stage 2, 3 or 4? What term do we give to the situation or type of climax that develops?

(c) Give an approximate time scale for the transition from stage 1 to stage 4.

(d) What term(s) do we give to the situation that develops through these stages?

(e) Which of the stages do you think would show greatest species diversity? Give reasons to support your answer. What happens to the profile of animal species in these different stages?

(f) In stage 4 (woodland), what differences might you expect in the ground flora at different times of year? (*Note:* assume this is a deciduous woodland in a temperate climate.)

3B.5 The diagrams below represent two other situations where a succession may occur. You may be able to think of others, based on your own field experience.

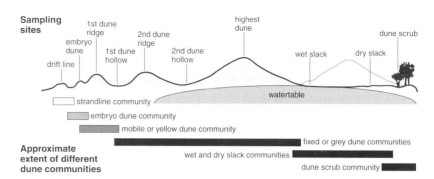

Sampling sites

drift line | embryo dune | 1st dune ridge | 1st dune hollow | 2nd dune ridge | 2nd dune hollow | highest dune | wet slack | dry slack | dune scrub

watertable

◻ strandline community
▨ embryo dune community
▨ mobile or yellow dune community

Approximate extent of different dune communities

■ fixed or grey dune communities
wet and dry slack communities ■
dune scrub community ■

Unit 5B

For each of these (diagrams or your chosen example), describe how it represents a succession and try to identify the different stages. Describe what you would expect to be characteristic of the climax community. Then suggest what could *prevent* the location reaching a climax community.

Populations

3B.6 As you work through this section on **populations**, make sure you try to relate the situations, including the definitions, to real examples as this will help to improve your understanding.

(a) Check your definitions (but with a variation) – this time, match relevant terms to the descriptions given.

    (i) *populations or communities?*

    A group of individuals in a species is a ....

    'All' organisms in a habitat (can be all organisms, all plants, a particular group of animals and / or plants) is described as a ....

    (ii) *numbers or density?*

    Population size refers to the .... in a population.

    Population .... refers to numbers in a given area or volume.

(b) Look next at population numbers and how and why they may change over time.

    Sketch a curve to show typical growth for a microbial population (say bacteria or yeast) inoculated into a fresh culture solution. What factors might limit numbers (or growth of the population)?

(c) Suppose a small number of organisms (e.g. a terrestrial animal) is introduced into a 'new' location and allowed to colonise the area. Predict what might happen to the population numbers of this animal and give a suggested *time* scale. What factors are likely to contribute to the success (or otherwise) of a species colonising a new area?

(d) We use a number of terms to describe the interactions between theoretical growth of a population and the real situation. The graph (left) gives a fairly realistic representation of population numbers over a period of time.

    Identify which numbers represent the words (terms) in the following list, two of which are for the axes of the graph.

    List of terms: *carrying capacity; density; environmental resistance; population curve; potential population curve; time;*

(e) Give a definition in words for *carrying capacity* and *environmental resistance*.

(f) Look next at *physical* factors that might affect population numbers.

Helpful hints

You could apply this to events that could occur in a river, just below the outflow of a polluting effluent (say from a sewage spill, or farm slurry).

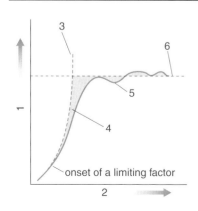

onset of a limiting factor

Helpful hints

To help you, think of some specific examples. Look at a population of unicellular algae, say in a lake, as a useful first example.

As examples here (g), you can refer to agricultural practices relating to weed control in crops and spacing of crop plants.

Population numbers can fluctuate in situations which depend on interactions between a predator and its prey (h).

How would numbers of unicellular algae in a lake be affected by changes in nutrients in the water? What situation might lead to a population explosion? How might numbers be affected by changes in light and in temperature? Give examples of when these may vary and how this would affect population numbers.

(g) Organisms within a community interact with each other in different ways and so can affect numbers of the populations. How does *intra*specific competition differ from *inter*specific competition? Which example (see margin box) fits with which type of competition and what do the different plants compete for?

As well as these two situations, think of an example of each type of competition using animals rather than plants.

(h) Give a definition of the terms *predator* and *prey*. Then sketch a graph to illustrate typical fluctuations in numbers which may occur between predator and prey. What sort of link can you make between this graph and methods used to control insect pests in crop plants?

(i) We have looked at different ways within a community that may lead to changes in numbers of a particular population of organisms. What *other* factors can result in changes in population numbers?

3B.7 (a) The table below gives some features or consequences of using chemical methods or biological methods to **control insect pests**. In the last two columns place a tick (✓) if the statement is correct for the control method or a cross (✗) if it is inappropriate.

Feature of control method	Chemical	Biological
Action general – can kill wide range of insects		
Action often specific – targets particular species or group of species		
Often does not completely eliminate the pest		
Toxicity may affect other (non-pest) animal species		
Acts relatively slowly		
May persist over long period – spread through food web etc		
Effectiveness may diminish as numbers of resistant strains increase in the population		

(b) In biological control methods, why might the time of release be critical? Give an example in which you relate release of the control organism to events such as life cycle of the pest or environmental factors. Why might it be counter-productive if the control organism completely eliminates the pest?

(c) Compare the relative costs of using chemical or biological control methods – not in terms of actual money, but by trying to weigh up some of the factors which would contribute to costs and which the farmer would need to consider when choosing the control method.

(d) Encouragement of natural predators can make a contribution to controlling numbers of a pest population. Find out about each of the following agricultural practices and suggest how they encourage natural predators.

List of agricultural practices: beetle banks; Phacelia tanacetifolia (*Note:* has blue flowers which attract hoverflies); *intercropping;*

(e) List some other ways that can be exploited to encourage natural predators.

(f) Use an example to explain what is meant by integrated pest management (IPM).

(g) Which of the 'features' in the table (Question 3B.7a) is effectively describing the bioaccumulation of non-biodegradable toxins? Give at least *one* example of where this has occurred.

Conservation

3B.8 Make a brief list of what you consider are the *aims* of **conservation** then give some reasons why, as we go into the twenty-first century, conservation is considered to be important in our consideration of the environment and its future.

3B.9 Look back to Question 3B.4 which is on succession in grassland.

(a) How can such an area of grassland be managed to maintain a range of habitats and to encourage species diversity? Make it clear how you can link this management to a 'strategy' for conservation.

(b) Apply the same principles to a freshwater situation, such as a pond or water channel (say between fields). How can this lead to a succession? What management strategies can you suggest to maintain a range of habitats? What effect might this have on species diversity?

3B.10 In Unit 3 (Human influences on the environment), coppicing in woodland is studied as an example of management for sustainability.

Now consider coppicing as an example of management for conservation. How far does this help achieve any of the aims as set out in Question 3B.8.

3B.11 Deliberate use of fire is also exploited as a management technique. What sort of habitat(s) is this used in and why?

3B.12 In your answer to Question 3B.8 you might have suggested that change of land use to farming leads to loss of 'natural' habitats.

(a) Which of the following systems is likely to sustain greater species diversity – small-scale intercropping or large-scale monoculture? Give a brief description of each to help explain your answer.

(b) How far can we describe any habitat as 'natural'?

(c) Make a list of practices adopted *within* farmland that can encourage species diversity.

> **Helpful hints**
>
> This question (3B.8) is really to start you thinking. You may come back to it and revise your initial list as you work through the rest of the questions.

> Even though you may not have studied such a habitat, think about what is likely to happen over time to the open water and why.

> **Helpful hints**
>
> In (c), you may select examples from within a UK farmland, or a farming system elsewhere. Whatever you choose, try to relate it to a real farming system, not just a hypothetical situation.

Helpful hints

We have now looked at conservation and management strategies in particular habitats – grassland, coppiced woodland, ponds or waterways. To encourage adoption or implementation of desired conservation measures often requires legislation.

Mark allocations are given for each part of the questions and the answers are given on pages 117–18.

3B.13 List some points from the EU Habitats Directive concerning the conservation of natural habitats and of wild fauna and flora and of Natura 2000. Then argue a case for and against legislation with respect to conservation strategies. Say how far you think such legislation can be successful.

Practice questions

1   A survey was carried out on a rocky sea shore to determine the distribution of marine molluscs, *Littorina saxatilis* (the rough periwinkle) and *L. littorea* (the common periwinkle). Both species are primary consumers. A profile of the rocky shore is shown on the diagram below. At low water mark, the shore is covered by sea water most of the time. The sea reaches high water mark twice each day.

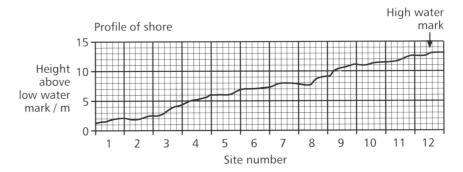

The sites of sampling were 10 metres apart, starting at the low water mark. The distributions were assessed by means of an abundance scale with 5 representing the greatest abundance. The results are shown as bar charts in the diagrams below.

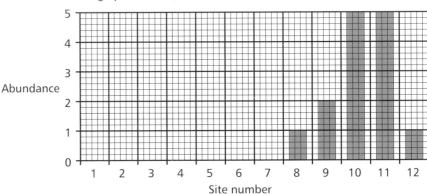

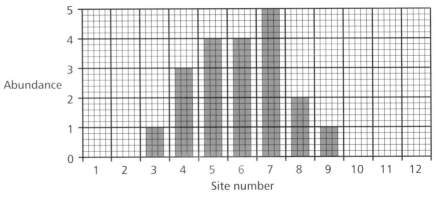

(a) Compare the distribution and abundance of these two species on this rocky shore. **(3 marks)**

(b) Suggest which of the two species is likely to be more tolerant of desiccation. Explain your answer. **(2 marks)**

(c) Suggest **two** factors, other than desiccation, which might account for the difference in distribution of the two species. **(2 marks)**

(Total 7 marks)
(Edexcel 6042, B2, June 1998, Q.5)

2   An insecticide (methyl parathion) was used in Texas in an attempt to control the tobacco budworm, *Heliothis virescens*. The effectiveness of the insecticide was monitored over a four year period, from 1967 to 1970. The insecticide was applied at three different concentrations (0.09 kg ha⁻¹, 0.27 kg ha⁻¹ and 0.54 kg ha⁻¹). The percentage of tobacco budworms killed each year at each concentration was recorded. The results are shown in the table below. No data are available for 1969.

Concentration of insecticide / kg ha⁻¹	Percentage of budworms killed each year		
	1967	1968	1970
0.09	99.9	70	20
0.27	Not used	90	40
0.54	Not used	90	50

(a) Describe how the effectiveness of the insecticide changed during the four year period. **(2 marks)**

(b) Suggest an explanation for this change in the effectiveness of the insecticide. **(4 marks)**

(c) Bioaccumulation of the toxin is an environmental problem which can occur as a result of the use of some insecticides.

(i) Describe how bioaccumulation occurs. **(3 marks)**

(ii) Suggest **two** ways in which insecticides may be designed to avoid the problems resulting from bioaccumulation. **(2 marks)**

(Total 11 marks)
(Edexcel 6042, B2, June 1998, Q.6)

3   Avocets and gulls are both birds that breed in coastal ecosystems. Gulls can feed on the eggs and young of the avocets.

A study was made of the relationship between the numbers of these two species in a particular coastal area. The numbers of birds of each species were recorded at two-yearly intervals over a period of thirty years between 1951 and 1981.

In 1965 some nests containing gulls' eggs were removed from the area as a management technique attempting to maintain both species within this coastal area. The changes in the numbers of each species are shown in the graphs below.

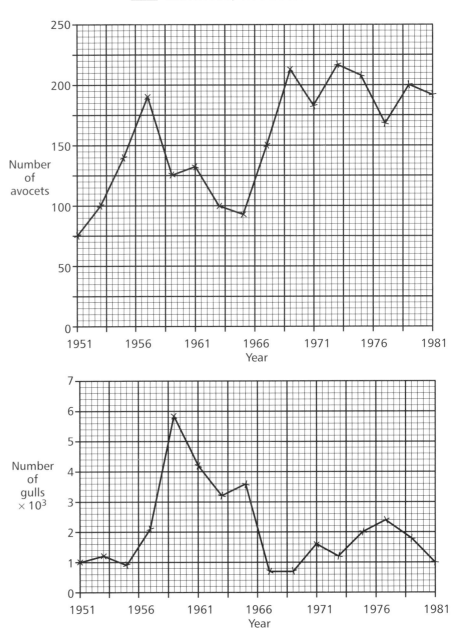

(a) Describe and explain the changes in the numbers of gulls and of avocets between 1951 and 1961. **(3 marks)**

(b) (i) Gulls' nests and eggs were removed in 1965. Calculate the percentage decrease in the number of gulls over the period 1965 to 1967. Show your working. **(3 marks)**

(ii) Describe how the decrease in the number of gulls affected the avocet population between 1965 and 1971. **(2 marks)**

(iii) Suggest **two** reasons for the changes in numbers of the avocets after 1971. **(2 marks)**

(c) Name **one** other management technique and describe its effect on the ecosystem in which it is used. **(3 marks)**

(Total 13 marks)

(Edexcel 6042, B2, January 1997, Q.6)

Topic 4B

Genetics and evolution

 Introduction

An understanding of genetics and evolution is an important part of studies in Biology. This topic builds on a number of concepts already introduced into the specification in earlier units and it would be useful to ensure that you have a good general understanding of the following before you begin your learning:

- **From Unit 1:**
  - the structure of nucleic acids
  - the nature of the genetic code
  - protein synthesis
  - the structure of chromosomes.
- **From Unit 2:**
  - adaptations to the environment
  - the process of sexual reproduction
  - the significance of meiosis for genetic variation.

As this topic is quite extensive, it is probably helpful to break it into smaller sections using the headings in the specification. This has been done for you in the checklist.

 Checklist of things to know and understand

Before attempting to answer any of the questions, check that you know and understand the following:

Genes and alleles

- ❏ gene expression and the environmental influences on gene expression
- ❏ the process of monohybrid inheritance
- ❏ the terms *genotype* and *phenotype*, *homozygote* and *heterozygote*, *dominance* and *codominance*
- ❏ the term *multiple alleles*, illustrated by the ABO blood-group system
- ❏ the inheritance of two non-interacting unlinked genes
- ❏ autosomal linkage and recombination in relation to the events of meiosis
- ❏ gene interaction between two unlinked genes
- ❏ sex determination in humans

Variation

❑ the nature of continuous and discontinuous variation

❑ that single gene inheritance is associated with discontinuous variation and polygenic inheritance is associated with continuous variation

Sources of new inherited variation

❑ the significance of meiosis and random fertilisation in sexual reproduction

❑ that crossing-over leads to recombinant chromosomes

❑ the significance of mutations

❑ the effects of chemical mutagens and radiation

❑ the nature of point mutations, as illustrated by base deletions, insertions and substitutions

❑ the effect of point mutation on amino acid sequences, as illustrated by sickle-cell anaemia in humans

❑ the nature of chromosome mutations, as illustrated by translocation

❑ that non-disjunction can lead to polysomy and polyploidy

Environmental change and evolution

❑ the process of natural selection

❑ that selection pressures change allele frequency in the population

❑ the processes of stabilising, directional and disruptive selection

❑ that isolating mechanisms lead to the divergence of gene pools

❑ geographical and behavioural isolating mechanisms

❑ the difference between *sympatric* and *allopatric* speciation

Gene technology

❑ the roles of reverse transcriptase, endonucleases and DNA ligase in the manipulation of DNA

❑ the processes involved in DNA insertion into a host cell and the multiplication of the host cell

❑ the use of marker genes to indicate that new genes have been incorporated into host cells

❑ how proteins are synthesised within host cells, as illustrated by the introduction of new genes into plants using the bacterium *Agrobacterium tumefaciens*

❑ the production of chymosin from genetically modified yeast and its use as a substitute for mammalian rennin in the dairy industry

❑ the potential of genetically modified organisms, as illustrated by crop plants and the development of pharmaceutical products

❑ the social, ethical and economic implications of the development of genetically modified organisms

❑ how the polymerase chain reaction (PCR) amplifies genetic material

❑ the process of genetic fingerprinting and its use as a diagnostic tool.

Practicals

You are expected to have carried out practical work to investigate:

☐ *the principles of inheritance.*

Practical work – Helpful hints

You are expected to have carried out a breeding experiment to illustrate the principles of inheritance. You could, for example, set up a monohybrid cross between two different strains of *Drosophila*, the fruit fly, such as a cross between normal-winged (wild-type) flies and vestigial-winged flies.

- You must start with cultures of pure-breeding wild-type and vestigial-winged flies.
- Set up a cross between virgin female wild-type flies and male vestigial-winged flies.
- Incubate the culture at 25 °C. After one week, larvae should be seen in the culture and the parent flies are removed.
- When the F_1 flies emerge, anaesthetise them and record their phenotypes. Set up new crosses with these F_1 flies.
- After one week, remove the F_1 flies, then leave the culture for the F_2 flies to emerge.
- Record the phenotypes of the F_2 flies.
- You could also set up a test cross with flies from the F_1 and pure-breeding flies with the homozygous recessive genotype.
- Apply a chi-squared test to the results of this experiment to investigate whether your results differ significantly from the expected results.
- Other organisms which are suitable for investigations in genetics include the flour beetle (*Tribolium castaneum*) and rapid cycling brassicas (*Brassica rapa*).

 Testing your knowledge and understanding

To test your knowledge and understanding of genetics and evolution, try answering the following questions.

Genes and alleles

4B.1 Different versions of an allele can exist in a population and these are termed *genes*. Is this statement true or false?

4B.2 For a gene given the symbol B, name the alleles present in:

(a) a heterozygous organism;

(b) a homozygous recessive organism.

4B.3 Fur colour in guinea pigs can be either black or brown. Does this statement describe the phenotype or the genotype of guinea pigs?

4B.4 Give an example of a multiple allele system in humans.

4B.5 Define the term *autosomal linkage*.

4B.6 In humans, are males or females the heterogametic sex?

 The answers to the numbered questions are on page 118.

Helpful hints

The terms *gene* and *allele* are often confused. Make sure that you know and understand the definition of each term.

Questions asking you to define terms are common in examinations. For each topic you study, it would be useful to make a glossary of terms. You might also find it helpful to refer to an A-level Biology dictionary.

Variation

4B.7 Give an example of a human characteristic that varies:
   (a) continuously;
   (b) discontinuously.

4B.8 Which of the statements (a) and (b) below refers to continuous variation and which refers to discontinuous variation?
   (a) Characteristic controlled by one or two genes, with little environmental influence.
   (b) Characteristic controlled by many genes, with considerable environmental influence.

Sources of new inherited variation

4B.9 Which type of cell division is important for increasing genetic variation?

4B.10 Why are mutations particularly important for increasing genetic variation?

4B.11 Give one example of:
   (a) a chemical mutagen;
   (b) a physical mutagen.

4B.12 What are the *three* possible types of point mutations?

Environmental change and evolution

4B.13 What is meant by the term *translocation*?

4B.14 Identify the three types of selection from the descriptions below.
   (a) This type of selection favours the mean of the population.
   (b) This type of selection favours one extreme of the population.
   (c) This type of selection favours both extremes of the population.

4B.15 What is the term used to describe the process by which new species are formed?

Gene technology

4B.16 What are the two types of isolating mechanisms that can lead to reproductive isolation?

4B.17 Identify the enzymes used in gene technology from their descriptions below:
   (a) an enzyme which cuts DNA molecules at specific points;
   (b) an enzyme which joins together pieces of DNA;
   (c) an enzyme which forms DNA from an RNA template.

4B.18 What is a plasmid?

4B.19 Why are plasmids useful in gene technology?

4B.20 Give *two* examples of proteins that are manufactured using recombinant DNA technology.

Helpful hints

A useful revision aid for point mutations would be to produce a table listing: the type of mutation, a description plus example and the effect of the mutation.

You could construct flow diagrams to show how: (a) a base substitution mutation can lead to sickle-cell anaemia; (b) non-disjunction during meiosis can lead to Down's syndrome.

It would be useful to be able to identify these three types of selection from graphs showing changes in the relative frequency of different phenotypes of an organism.

You could devise a table, using the headings *social*, *ethical* and *economic*, to summarise the implications (positive and negative) of genetically modified organisms.

A useful revision activity would be to make a large poster, describing the use of genetic fingerprinting as a diagnostic tool.

Mark allocations are given for each part of the questions and the answers are given on pages 118–19.

 Practice questions

Genes and alleles

1   One example of a multiple-allele system is seen in the gene for the determination of coat colour in rabbits, where there are four alleles:

● agouti (**A**) with yellow and black fur

● chinchilla (**C**) with grey fur

● himalayan (**H**) with white fur, except for black nose, ears, feet and tail

● albino (**W**) with no pigment, the fur is pure white

The dominance sequence is **A** > **C** > **H** > **W**. Complete the table below to show the possible genotypes of the four different phenotypes.

Phenotype	Genotype
Agouti	
Chinchilla	
Himalayan	
Albino	

(Total 4 marks)
(New question)

2   Maize cobs may have purple or red grains. This character is controlled by a single pair of alleles. The dominant allele **A** gives a purple colour and the recessive allele **a** gives a red colour.

(a) In an experiment, a heterozygous plant is crossed with a maize plant homozygous for allele **a**. state the genotype of these two plants.   **(1 mark)**

COPLAND COMMUNITY SCHOOL
FACULTY OF SCIENCE

15

BIOLOGY

(b) Grain colour is also affected by a second pair of alleles. The presence of the dominant allele **E** allows the purple or red colour to develop, but in the homozygous recessive (**ee**) no colour will develop (despite the presence of alleles **A** or **a**) and the grain will be white. A plant of genotype **AAEE** is crossed with a plant of genotype **aaee**.

(i) State the genotype and phenotype of the offspring produced as a result of this cross.   **(2 marks)**

(ii) The plants of the offspring are allowed to self-fertilise. Draw a genetic diagram to show the possible genotypes produced as a result of this cross.   **(3 marks)**

(iii) Predict the phenotypic ratio that would be obtained from this cross.   **(3 marks)**

(iv) Which genotypes, if allowed to self-fertilise, would produce pure-breeding lines containing white grains?   **(3 marks)**

(Total 12 marks)
(Edexcel 6041, B/HB1, January 1996, Q.6)

Variation

3   Distinguish between each of the following pairs of terms, illustrating your answer with suitable examples.

    (a) Monohybrid inheritance and dihybrid inheritance. **(3 marks)**

    (b) Continuous variation and discontinuous variation. **(3 marks)**

                                              (Total 6 marks)

                                  (Edexcel 6041, B/HB1, January 2001, Q.4)

Sources of new inherited variation

4   Distinguish between each of the following pairs of terms.

    (a) Postzygotic and prezygotic reproductive isolation. **(3 marks)**

    (b) Sympatric and allopatric speciation. **(3 marks)**

                                              (Total 6 marks)

                                  (Edexcel 6042, B2, January 1999, Q.4)

5   T-toxin is an insecticidal substance which is produced by a species of bacterium, *Bacillus thuringiensis*. Using the techniques of gene technology, it is possible to isolate the gene responsible for the production of T-toxin from bacteria, and to transfer this into tomato plants. An outline of this procedure is shown in the diagram below.

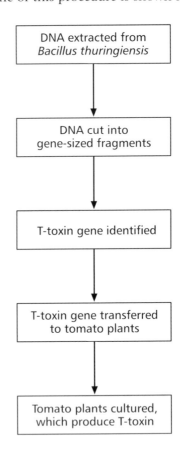

    (a) **(i)** Name the enzyme which can be used to cut DNA into 'gene-sized fragments'. **(1 mark)**

          (ii) Suggest how the gene for T-toxin could be identified. **(2 marks)**

    (b) Suggest how the gene for T-toxin could be inserted into tomato cells. **(2 marks)**

(c) T-toxin is a protein. Describe the process by which genetically modified tomato plants will synthesise T-toxin. **(4 marks)**

(d) Suggest **one** possible advantage and *one* possible disadvantage of growing tomato plants into which the gene for T-toxin has been inserted. **(2 marks)**

(Total 11 marks)

(Edexcel 6041, B/HB1, January 1999, Q.6)

6   **(a)** Distinguish between the terms *gene* and *allele*. **(3 marks)**

(b) The diagram below shows a family tree in which the blood group phenotypes are shown for some individuals.

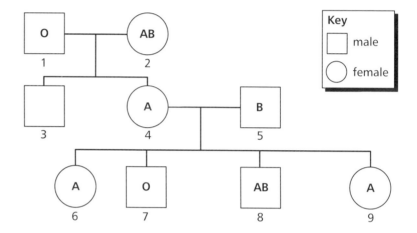

(i) Using the symbols I^A, I^B and I^O to represent the alleles, indicate the genotypes of the following people: 1, 2, 4, 5, 6. **(5 marks)**

(ii) State the possible blood groups of person 3. Explain your answer. **(3 marks)**

(Total 11 marks)

(Edexcel 6041, B/HB1, January 1997, Q.7)

Mark allocations are given for each part of the questions and the answers are given on pages 126–8.

Assessment questions

1   An area of abandoned grassland was studied over a period of more than 100 years. The table below shows the changes which occurred in the plant communities and in the number and density of species of small birds.

Time since abandoned (years)	1–10	10–25	25–100	100+
Plant community	Grass	Shrubs	Pine trees	Mixed woodland
Number of species of small birds	2	8	15	19
Density (pairs of birds per 40 hectares)	27	123	113	233

(a) Suggest why the grass community is replaced by the shrub community. [2]

(b) Calculate the percentage change in the number of species of small birds as the plant community changes from shrubs to mixed woodland. Show your working. [2]

(c) Suggest reasons for the change in the number of small birds species as the community changes from shrubs to mixed woodland. [2]

(d) Suggest why the density of birds decreased when the shrub community was replaced by pine trees. [2]

(e) What name is given to the process by which communities change over time, as illustrated by this example? [1]

(f) The final climax vegetation can be stable for long periods of time. Suggest **two** ways in which this stable community may be disturbed. [2]

(Total 11 marks)
(Edexcel 6042, B2, January 2001, Q.7)

2   The flatworm *Dendrocoelum lacteum* lives in the shallow waters of lakes and ponds where it clings to the underside of rocks and aquatic vegetation. It is an efficient predator, feeding on small worms, snails and arthropods. One of its main food sources is the isopod crustacean, *Asellus meridianus*.

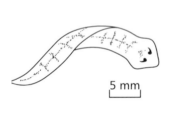

Dendrocoelum

Asellus

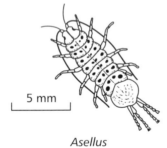

The numbers of *Dendrocoelum* and *Asellus* in a pond in Northern Europe were recorded over one year. The graph above shows the numbers of both species in a metre-square area of the pond between January and December.

(a) Define the term **population**. [2]

(b) Compare the changes in the numbers of Dendrocoelum and Asellus between January and December. [3]

(c) Suggest reasons for the changes in the numbers of both species between April and September. [3]

(d) Describe how you would estimate the population density of the flatworms in a pond using quadrats. [4]

(Total 12 marks)
(Edexcel 6042, B2, June 2001, Q.7)

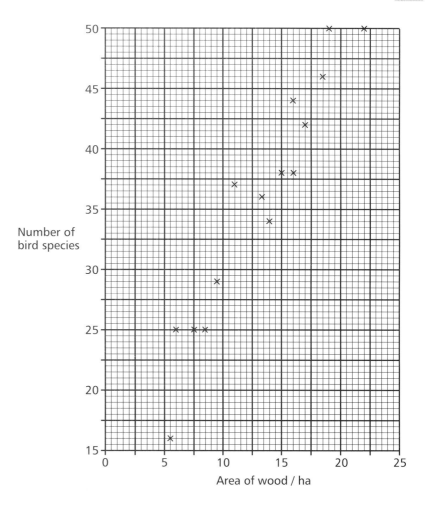

Number of bird species (y-axis, 15 to 50)

Area of wood / ha (x-axis, 0 to 25)

3 An investigation was carried out into the relationship between the size of a wood and the number of species of birds in it. Sixteen woods of different sizes were studied in the investigation. The size of each wood was measured in hectares and the number of different species observed in each wood was noted.

The results are shown in the scatter graph above.

(a) Suggest **two** precautions that should be taken to ensure that the results obtained from the different woods are comparable. Give a reason for each precaution. **(4 marks)**

(b) State **two** reasons why there is a general trend for the number of bird species to increase as the size of the wood increases. **(4 marks)**

(c) Conservation techniques may be used to maintain or increase the number of bird species in a woodland. In some woods, rides are created by clearing trees from 5 metre strips of land.

Suggest how this practice might increase the number of bird species. **(3 marks)**

(Total 11 marks)
(Edexcel 6042, B2, January 1999, Q.7)

4   (a)   One of the genes that controls coat colour in cats has its locus on the X chromosome. The alleles are O^o, giving orange coat colour, and O^b, giving black coat colour. The genotype O^oO^b produces a coat with a mixture of orange and black, known as tortoiseshell.

A female cat with orange coat colour was mated with a black male. The resulting offspring were tortoiseshell females and orange males.

       (i)   With reference to the alleles O^o and O^b, explain why all the male kittens from this cross had orange coats.   **[3]**

       (ii)   The black male cat was later mated with a tortoiseshell female. By mean of a genetic diagram, show the possible genotypes and phenotypes in the offspring of this cross.   **[4]**

   (b)   The expression of a particular phenotype may be influenced by environmental factors.

In an adult Siamese cat, the fur on the ears, face, feet and tail is coloured, due to the presence of a pigment. In Siamese cats, the production of this pigment is controlled by the action of the enzyme tyrosinase, which is temperature-dependent. The other regions of the body lack colour or the colour is very reduced,

Suggest why Siamese kittens have white fur when they are born and do not develop their characteristic markings until some days after their birth.   **[3]**

   (c)   Manx cats have no tails. The gene responsible for tail length in cats has two alleles, **M** (Manx, tailless) and **m** (tailed). The Manx condition is dominant.

When Manx cats are interbred and also when they are mated with tailed cats, some of the offspring are Manx and some have normal tails.

The average number of kittens born when two Manx cats are interbred is less than when two tailed cats are interbred, or when a Manx cat is mated with a tailed cat.

Suggest explanations for these observations, using genetic diagrams to illustrate your answer.   **[3]**

(Total 13 marks)
(Edexcel 6046, B6, June 2001, Q.3)

5   The table below refers to some of the major groups of living organisms (kingdoms), their characteristic features and examples of representative groups.

Complete the table by writing in the appropriate word or words in the boxes.

Kingdom	Characteristic feature	Representative group
	Lack envelope-bound organelles	
Protoctista	Possess envelope-bound organelles; often unicells or assemblages of similar cells	
Plantae		Angiosperms
	Non-photosynthetic organisms with multinucleate hyphae	Ascomycota, such as *Penicillium*

(Total 5 marks)
(New question)

6 The photosynthetic pigments present in a leaf extract were separated using thin-layer chromatography. The resulting chromatogram is shown below.

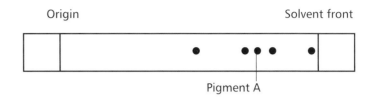

Origin   Solvent front

Pigment A

(a) The Rf of a pigment is calculated using the formula:

$$R_f \text{ of a pigment} = \frac{\text{Distance moved by pigment}}{\text{Distance from solvent front}}$$

The distance is measured to the centre of the pigment area.

Calculate the R_f of pigment A on the chromatogram. Show your working. **[2]**

(b) Identify pigment A using the information provide in the table below.

Name of pigment	R_f value
Carotene	0.96
Chlorophyll a	0.75
Chlorophyll b	0.70
Xanthophyll a	0.51

[1]

(c) Describe how this pigment is involved in photosynthesis. **[3]**

(Total 6 marks)
(Edexcel 6042, B2, June 2001, Q.4)

7 The diagram below is a summary of the processes that take place during the light-independent stage of photosynthesis in a palisade cell.

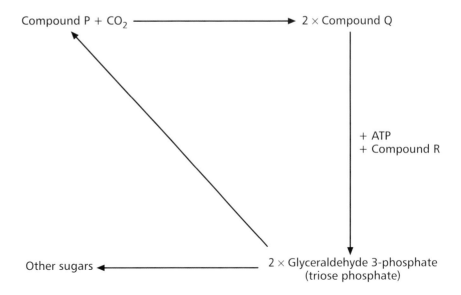

Compound P + CO_2 ⟶ 2 × Compound Q

+ ATP
+ Compound R

Other sugars ⟵ 2 × Glyceraldehyde 3-phosphate (triose phosphate)

(a) Name compounds P and Q and R. **[3]**

(b) Some of the glyceraldehyde 3-phosphate produced is recycled to form compound P as shown. What name is given to this whole cycle? **[1]**

(c) Some of the sugars produced are stored temporarily within the palisade cell. Name the form in which they are stored and state precisely where this store is located within the cell. **[2]**

(Total 6 marks)
(Edexcel 6042, B2, June 2001, Q.5)

8 An investigation was carried out into the effect of cytokinin on the growth of lateral (axillary) buds on the shoots of pea plants (*Pisum* sp.).

Three batches of pea seedlings were grown under identical conditions. The treatments of the three batches are described below;

Batch A The shoots were left intact.

Batch B The apical buds (tips) of the shoots were removed.

Batch C The shoots were left intact but a paste containing 2 ppm of cytokinin was applied to the lateral buds.

The lengths of all the lateral buds on the shoots in each batch were measured at the beginning of the investigation and the mean lateral bud length for each batch was calculated. The measurements were repeated after 3, 5 and 7 days for each batch of pea seedlings.

The results are shown in the table below.

Time / days	Mean lateral bud length		
	Batch A Intact shoots removed	Batch B Apical buds buds	Batch C Cytokinin on lateral
0	1.75	1.75	1.75
3	1.75	6.25	2.60
5	1.75	8.25	3.50
7	1.75	8.50	6.40

(a) Explain why the mean length of the lateral buds on the shoots in Batch A did not change. **[2]**

(b) Compare the results obtained for Batch B with those for Batch C, where cytokinin has been applied to the lateral buds. **[3]**

(c) Suggest an explanation for the effect of the cytokinin on the mean lateral bud length in Batch B. **[2]**

(d) Cytokinins are plant growth substances involved in chemical coordination in plants. Describe how plant growth substances differ from animal hormones. **[3]**

(Total 10 marks)
(Adapted from Edexcel 6046, B6 Synoptic, June 2001, Q.5)

5H Genetics, human evolution and biodiversity

Introduction

Unit 5H is the Biology (Human) pathway. If you are studying Biology go to page 41.

This unit contains aspects of Biology which bring together and extend topics covered in other units of the specification. The first topic is concerned with the transmission of genetic information from generation to generation and it is important to understand the nature of variation and inheritance. It is essential that you have a sound knowledge of the specific terms used and that you can explain the mechanisms of both monohybrid and dihybrid inheritance, linkage, gene interaction and sex determination. This topic includes reference to sources of new inherited variation and some aspects of natural selection. The manipulation of DNA and the introduction of new genes into organisms leads to a consideration of some of the applications of gene technology. The topics of human evolution and human populations are specialist areas and give you an opportunity to gain some understanding of how the human population has evolved and how it has grown. The biodiversity topic is similar to that of the Biology specification and requires the study of at least one habitat.

At all stages of your learning and revision of this unit, you should be prepared to review related topics in the AS units and in Unit 4. The written test for this unit contains a synoptic assessment, which will involve questions drawing on material from all units of the specification. For each topic, the related material from other units is highlighted for you.

This unit is divided into four topics:

1H Genetics and evolution
2H Human evolution
3H Human populations
4H Biodiversity

Topic 1H

Unit 5H

Genetics and evolution

Introduction

An understanding of genetics and evolution is an important part of studies in Biology. This topic builds on a number of concepts already introduced into the specification in earlier units and it would be useful to ensure that you have a good general understanding of the following before you begin your learning:

- **From Unit 1:**
  - the structure of nucleic acids
  - the nature of the genetic code
  - protein synthesis
  - the structure of chromosomes
- **From Unit 2:**
  - adaptations to the environment
  - the process of sexual reproduction
  - the significance of meiosis for genetic variation.

As this topic is quite extensive, it is probably helpful to break it into smaller sections using the headings in the specification. This has been done for you in the checklist.

Note that this specification contains an extra section on genetic counselling which does not appear in the Biology specification.

Checklist of things to know and understand

Before attempting to answer any of the questions, check that you know and understand the following:

Genes and alleles

❏ gene expression and the environmental influences on gene expression

❏ the process of monohybrid inheritance

❏ the terms *genotype* and *phenotype*, *homozygote* and *heterozygote*, *dominance* and *codominance*

❏ the term *multiple alleles*, illustrated by the ABO blood-group system

❏ the inheritance of two non-interacting unlinked genes

❏ autosomal linkage and recombination in relation to the events of meiosis

❏ gene interaction between two unlinked genes

❏ sex determination in humans

Variation

❑ the nature of continuous and discontinuous variation

❑ that single gene inheritance is associated with discontinuous variation and polygenic inheritance is associated with continuous variation

Sources of new inherited variation

❑ the significance of mutations

❑ the effects of chemical mutagens and radiation

❑ the nature of point mutations, as illustrated by base deletions, insertions and substitutions

❑ the effect of point mutation on amino acid sequences, as illustrated by sickle-cell anaemia in humans

❑ the nature of chromosome mutations, as illustrated by translocation

❑ that non-disjunction can lead to polysomy and polyploidy

Genetic counselling

❑ the risks of inherited diseases as determined from family history

❑ genetic screening and detection of fetal abnormalities by amniocentesis and chorionic villus sampling

❑ karyotyping

❑ the possible courses of action following the detection of pre-natal disorders, including treatment and termination of pregnancy

❑ the potential advantages and disadvantages of gene therapy

❑ the social, ethical and legal implications of genetic counselling

Environmental change and evolution

❑ the process of natural selection

❑ that selection pressures change allele frequency in the population

❑ the processes of stabilising, directional and disruptive selection

❑ that isolating mechanisms lead to the divergence of gene pools

❑ geographical and behavioural isolating mechanisms

❑ the difference between *sympatric* and *allopatric* speciation

Gene technology

❑ the roles of reverse transcriptase, endonucleases and DNA ligase in the manipulation of DNA

❑ the processes involved in DNA insertion into a host cell and the multiplication of the host cell

❑ the use of marker genes to indicate that new genes have been incorporated into host cells

❑ how proteins are synthesised within host cells, as illustrated by the introduction of new genes into plants using the bacterium *Agrobacterium tumefaciens*

☐ the production of chymosin from genetically modified yeast and its use as a substitute for mammalian rennin in the dairy industry

☐ the potential of genetically modified organisms, as illustrated by crop plants and the development of pharmaceutical products

☐ the social, ethical and economic implications of the development of genetically modified organisms

☐ how the polymerase chain reaction (PCR) amplifies genetic material

☐ the process of genetic fingerprinting and its use as a diagnostic tool.

Practicals

You are expected to have carried out practical work to investigate:

☐ *the principles of inheritance*

☐ *the preparation of a karyotype from a print of human metaphase chromosomes.*

 Practical work – Helpful hints

You are expected to have carried out a breeding experiment to demonstrate the principles of inheritance. You could, for example, set up a monohybrid cross between two different strains of *Drosophila*, the fruit fly, such as a cross between normal-winged (wild-type) flies and vestigial-winged flies.

● You must start with cultures of pure-breeding wild-type and vestigial-winged flies.

● Set up a cross between virgin female wild-type flies and male vestigial-winged flies.

● Incubate the culture at 25 °C. After one week, larvae should be seen in the culture and the parent flies are removed.

● When the F_1 flies emerge, anaesthetise them and record their phenotypes. Set up new crosses with these F_1 flies.

● After one week, remove the F_1 flies, then leave the culture for the F_2 flies to emerge.

● Record the phenotypes of the F_2 flies.

● You could also set up a test cross with flies from the F_1 and pure-breeding flies with the homozygous recessive genotype.

● Apply a chi-squared test to the results of this experiment to investigate whether your results differ significantly from the expected results.

● Other organisms which are suitable for investigations in genetics include the flour beetle (*Tribolium castaneum*) and rapid cycling brassicas (*Brassica rapa*).

You are also expected to have prepared a karyotype from a print of human metaphase chromosomes. A karyotype is a pictorial representation of all the chromosomes in a body cell of an individual. Cells are cultured so that they are actively dividing, then the cells are broken open and the chromosomes spread on a microscope slide. The chromosomes are stained so that parts of them form dark bands. The chromosomes are then photographed under a microscope and arranged in a specific order to form the karyotype. In humans, one system used for identifying and arranging the chromosomes is based on the following features:

◆ the overall size of the chromosome

◆ the position of the centromere

◆ the banding pattern.

- Carefully cut out each chromosome from the photograph supplied.
- Using the banding pattern for reference, organise the chromosomes on a sheet of paper, then stick them down to form the karyotype.
- Label your karyotype fully, including the numbers of each pair of autosomes and the X and Y chromosomes.
- You could prepare karyotypes for individuals with Down's syndrome and Klinefelter's syndrome.

 Testing your knowledge and understanding

The answers to the numbered questions are on pages 118 and 119.

To test your knowledge and understanding of genetics and evolution, try answering the following questions.

Sections on:

- genes and alleles
- variation
- sources of new inherited variation
- environmental change and evolution
- gene technology

are the same for Biology and Biology (Human). See pages 67–8.

Genetic counselling

1H.1 Name *two* techniques used in the detection of fetal abnormalities.

1H.2 What is meant by the term *karyotype*?

 Practice questions

Mark allocations are given for each part of the questions and the answers are given on pages 119–120.

See pages 69–71 for questions on:

- genes and alleles
- variation
- sources of new inherited variation.

Unit 5H

Genetic counselling

1   The photograph shows a karyotype of a boy with Down's syndrome.

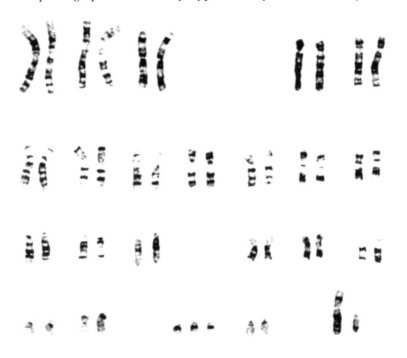

(a) State the number of chromosomes shown in this karyotype.

(1 mark)

(b) On a copy of the karyotype, label the Y chromosome. **(1 mark)**

(c) Down's syndrome is an example of polysomy, and may arise as a result of non-disjunction during the formation of a female gamete.

State what is meant by the term *polysomy*. **(1 mark)**

(d) Explain how fertilisation by normal sperm of a female gamete formed by non-disjunction may result in Down's syndrome.

(3 marks)

(Total 6 marks)

(Edexcel 6041, B/HB1, June 1999, Q.5)

Topic **2H**

Human evolution

Introduction

This topic considers humans as primates and the evolution of hominoids and *Homo*. It also examines the evidence for human evolution. Knowledge and understanding of the process of evolution covered in Topic 1H of this unit is therefore applicable to this section of the specification, although examples must relate to the evolution of humans.

Checklist of things to know and understand

Before attempting to answer any of the questions, check that you know and understand the following:

Humans as primates

❏ the range of form in primates, as illustrated by lemurs, Old and New World monkeys, apes and humans

Evidence for human evolution

❏ the phylogenetic relationships of hominoids and the evidence for human evolution (from comparative anatomy, fossils and geochronology and comparative physiology)

Hominoid evolution

❏ the divergence of apes and hominids from a common ancestor

❏ the possible influences of climate and habitat change on the evolution of hominid features

❏ features of the Australopithecines and their possible relationships to *Homo*

Evolution of Homo

❏ changes in the skeleton, skull and brain development; bipedalism and evolution of the hand

❏ the main features of *Homo habilis*, *Homo erectus*, *Homo sapiens* and Neanderthal Man

❏ cultural development of *Homo* spp. during the Palaeolithic

❏ the development of agriculture and settled communities, the domestication of animals during the Neolithic.

 Testing your knowledge and understanding

 The answers to the numbered questions are on page 120.

To test your knowledge and understanding of human evolution, try answering the following questions.

Humans as primates

2H.1 State *two* differences, excluding overall size, between the skull of a New World monkey and that of a chimpanzee.

2H.2 Which group of primates fits the following description?

Have a tail which is not prehensile. Most digits have nails, but their second toes have a claw which is used for grooming.

Evidence for human evolution

2H.3 Outline *three* molecular methods for investigating human evolution.

2H.4 Outline *two* non-molecular methods for investigating human evolution.

2H.5 Explain what is meant by the term *geochronology*.

Hominoid evolution

2H.6 Put the following terms in order, starting with the earliest and finishing with the most recent stage of human evolution: *Homo habilis*; *Homo sapiens*; Hominoids; Australopithecines; *Homo erectus*.

Helpful hints

You could construct a dichotomous key to identify members of the main groups within the primates (lemurs, Old and New World monkeys, apes and humans).

It might be useful to produce a spider diagram summarising the different types of evidence relating to human evolution.

2H.7 State *two* differences between the skull of Neanderthal man and that of a modern human.

Evolution of Homo

2H.8 State *three* differences between the cultural developments associated with *Homo habilis* and *Homo sapiens*.

2H.9 **(a)** Name the cultural period when the domestication of animals first took place.

(b) Give the approximate date, in years before the present, of the start of this cultural period.

2H.10 Explain how the development and use of flint knapping could have contributed to the way of life of *Homo sapiens* in the Upper Palaeolithic era.

 Practice questions

Mark allocations are given for each part of the questions and the answers are given on page 120.

1 The diagrams A, B, and C below show three reconstructed fossil skulls of the genus *Homo*.

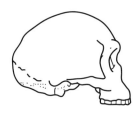 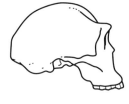 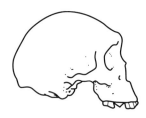

Skull A   Skull B   Skull C

(a) Arrange the letters in order with the earliest skull first and the most modern skull last **(1 mark)**

(b) Give **three** features shown by these skulls which support your answer in (a). **(3 marks)**

(Total 4 marks)
(Edexcel 6049, HB2, June 1998, Q.1)

2 The closeness of the relationship between different hominoids can be estimated by comparing their blood sera using a precipitation test. Antibodies precipitate proteins in the blood serum.

In such a test, human serum is injected into another animal, such as a rabbit. The rabbit responds by producing antibodies. When these antibodies are isolated and mixed with human blood serum, precipitation occurs. If, however, serum from a different hominoid species is used, less precipitation occurs. Species which are closely related to humans produce almost as much precipitation, whereas more distantly related species produce a smaller precipitate.

The results of a blood serum test on a range of hominoids are given below, expressed as percentages.

Human	100%
Chimpanzee	97%
Gorilla	92%
Gibbon	79%
Baboon	75%

(a) Antibodies are specific in relation to the molecules with which they interact. In this context, give the meaning of the term *specific*. **(1 mark)**

(b) What do the results indicate about the similarity to human blood proteins of the proteins in the blood of the baboon? Explain your answer. **(2 marks)**

(c) The diagrams below show two possible genealogical trees linking four of the above species.

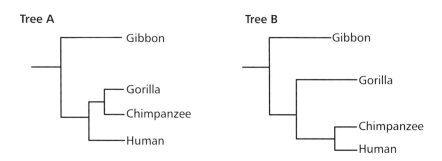

Tree A

- Gibbon
- Gorilla
- Chimpanzee
- Human

Tree B

- Gibbon
- Gorilla
- Chimpanzee
- Human

   (i) Which of the trees, A or B, is best supported by the evidence from the serum precipitation test? Give a reason for your answer. **(2 marks)**

   (ii) On the tree you have chosen in your answer to (i) add a branch which best shows the relationship of the baboon to the other species as indicated by the precipitation test. **(1 mark)**

(d) Explain why similar protein structures may indicate a close genetic relationship between the two species. **(2 marks)**

(e) An alternative method of comparing the proteins of related species is to find the amino acid sequences of specific protein molecules such as haemoglobin. Explain why this may be considered a better method than serum precipitation. **(3 marks)**

(Total 11 marks)
(Edexcel 6049, HB2, June 1997, Q.7)

3   Read through the following passage about the development of human social life during the Palaeolithic period, then fill in the most appropriate word or words to complete the passage.

The earliest species to be included in the genus *Homo*, *H.............*, lived during the Palaeolithic period, about 2 million years ago. The limited remains of this species are associated with crude stone tools. According to the fossil record, at about 1.5 million years BP a new species, *H.............*, appears. This species seems to have been a hunter using stone tools and .... The modern species, *Homo sapiens*, first occurs in deposits dating from .... The stone tools used during this part of the Palaeolithic are more highly developed and other artefacts of this period include pottery, basketry and .... It is now thought that two subspecies existed during this period. One of these, represented by Cro-Magnon man, was probably a direct ancestor of modern humans. The other, known as ...., probably became extinct.

(Total 6 marks)
(Edexcel 6049, HB2, June 2001, Q.2)

Topic 3H

Human populations

Introduction

This topic looks at human population trends on three levels: the global implications of world population trends, the factors that affect the growth of populations, past and present, in developed and developing countries, and the individuals that make up the populations. You will need to be familiar with the terminology used and be aware of the factors that lead to changes in birth and death rates.

Your learning and revision of this topic should be linked to Unit 3 of the AS, where human influences on the environment were considered, and to Topic 2H of this unit.

Checklist of things to know and understand

Before attempting to answer any of the questions, check that you know and understand the following:

❏ how world population has changed over time and the implications of these trends

❏ the factors that affect growth and size of human populations

❏ the factors which can cause variations in fertility and changes in birth and death rates

❏ what is meant by and how to interpret demographic trends

❏ how to interpret growth curves and population pyramids in countries with stable, increasing and declining populations.

 The answers to the numbered questions are on pages 120–2.

Helpful hints

Note that the scale here (in the graph) is in *millions* rather than *billions*.

Helpful hints

In (a), even if you cannot give the actual events you can try to think of a range of possible events which might have contributed to this.

Testing your knowledge and understanding

To test your knowledge and understanding of human populations, try answering the following questions.

3H.1 The graph below shows the annual addition to the population over the years 1950 to 1998.

(a) Suggest an explanation for the dip in addition to the world population shown in the early 1960s.

(b) At what other times in the twentieth century might there have been a similar dip in addition to the world population, and why?

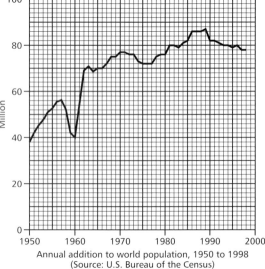

Annual addition to world population, 1950 to 1998
(Source: U.S. Bureau of the Census)

(c) Which countries (or group of countries) are likely to be contributing most to this annual addition to the world population in the period shown in this graph? One or two hundred years ago, which countries (or group of countries) contributed most to the annual addition to the world population?

(d) Which country in the world currently has the largest population and roughly what is its population (in billions)? What is the current population of the UK (or of your country if you are not a UK resident)?

(e) Check your definitions: *birth rate* and *death rate*; *high stationary*, *expanding* and *low stationary* in the demographic model of population growth.

3H.2 (a) Draw a population pyramid.

(b) Explain why the pyramid is narrower at the top, then decide whether the one you have drawn would be typical of a developed or of a developing country. Which of the following does it represent – a *stable*, *increasing* or *declining* population? Sketch curves to represent the two other types of population. What other 'events' can alter the shape of the pyramid?

3H.3 (a) Check your definitions: *fertility* and *fecundity*; *crude birth rate* and *crude death rate*.

(b) How are the terms *general fertility rate* and *total fertility rate* a useful refinement on the term *crude birth rate*?

(c) Work through the following list and make brief comments on how these may affect fertility rates in a population. In each case, try to quote an example which illustrates the points you are making.

List: *proportion of women in marriage; age of marriage; family size.*

(c) How is family size controlled in 'traditional' families and how is it controlled in modern societies? Use world-wide examples, not just from your own country.

3H.4 The graph below shows changes in sperm count in American and European men between 1938 and 1990.

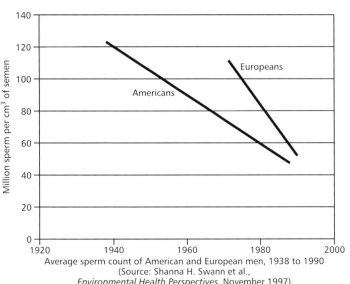

Average sperm count of American and European men, 1938 to 1990
(Source: Shanna H. Swann et al.,
Environmental Health Perspectives, November 1997)

Helpful hints

In (e), you would find it helpful to sketch a graph and annotate the curves with the meanings of these terms.

Think about how you organise the horizontal axis, what 'units' you use there, what the horizontal bars represent and what goes into the two halves of the pyramid.

Helpful hints

The term 'marriage' is also used to cover those living in a stable relationship with a partner

---

In (d), try to give examples of some government measures to control family size, with respect both to reducing the number of births and to encouraging increases in the number of births.

Helpful hints

Pay particular attention to events in recent years that may have led to these decreases.

 Mark allocations are given for each part of the questions and the answers are given on page 122.

(a) Suggest possible *reasons* for these downward trends in sperm count. Then suggest possible *consequences* of these trends in terms of population numbers.

(b) For what *physiological* reasons (other than reduced sperm count) may couples experience reduced fertility (subfertility) or failure to conceive and give birth to a child?

(c) List some of the ways that couples can be helped to overcome subfertility (or infertility).

 Practice questions

1   The Total Fertility Rate (TFR) for a country is a measure of the average number of children expected to be born to a woman. It assumes that the woman lives to the end of her child-bearing age, and that she produces children at the same rate as other women did in the year of the calculation. A TFR of 2.1 to 2.5 is considered to be the level needed for natural replacement of the population.

The table below gives the TFR for ten countries.

Country	Total Fertility Rate (TFR)
Bulgaria	2.1
Costa Rica	3.5
Finland	1.6
Jordan	6.5
Nepal	2.5
Republic of Korea	2.5
Romania	2.4
Thailand	3.8
United Kingdom	1.8
United States of America	1.8

(a) (i) Using this information, select **two** countries which appear to have an increasing population and **two** countries which appear to have a stable population. **(2 marks)**

(ii) Explain why there are variations in TFR between countries. **(3 marks)**

(b) State **one** factor, other than fertility, which would affect the growth of the population of a country. **(1 mark)**

(Total 6 marks)
(Edexcel 6049, HB2, January 1999, Q.4)

2   It is possible to predict the future population pyramid of a country, based on the current population pyramid together with the birth rate and life expectancy.

(a) Define the term **birth rate**. **(2 marks)**

(b) The diagrams below show the population pyramid of a developing country in 1990 and the predicted population pyramid for the same country in 2020.

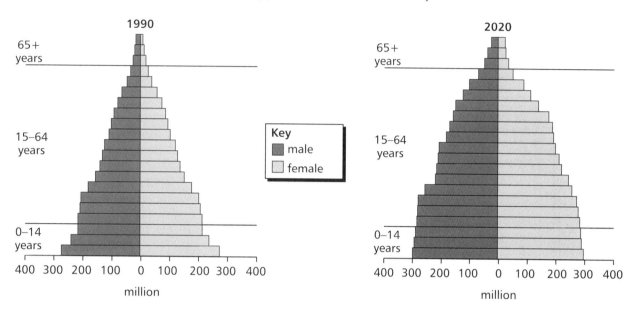

Describe and explain **two** differences between the population pyramids of 1990 and 2020. **(4 marks)**

(Total 6 marks)

(Edexcel 6049, HB2, June 2001, Q.4)

3 The diagrams below show population pyramids for Great Britain for the years 1891 and 1947.

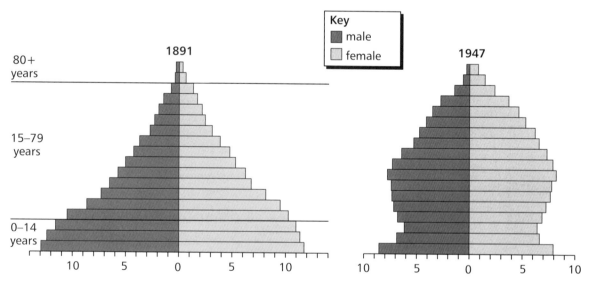

(a) State which of these pyramids is most likely to show a rapidly increasing population. Give a reason for your answer. **(2 marks)**

(b) Compare the composition of the populations aged 30 and above in 1891 with the same age range in 1947. Suggest why there are differences between these populations. **(4 marks)**

(Total 6 marks)

(Edexcel 6049, HB2, June 1998, Q.2)

Unit 5H

Unit 5H

Topic 4H

Biodiversity

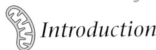
Introduction

This is a broad topic, encompassing the distribution of plants and animals, succession, the control of insect pests and conservation. It is essentially similar to Topic 5B of the Biology specification (Unit 5B), except that you do not need to study Classification and only part of the section on Populations has been included.

You are required to have an understanding of the effects of abiotic and biotic factors on the distribution of organisms in terrestrial and aquatic habitats. This can be achieved through practical investigations that involve the use of qualitative and quantitative sampling techniques. It is important to relate your revision of this topic to your practical work and to a specific habitat, as you should be familiar with the organisms in the habitat you have studied and with the techniques you have used. A knowledge of the details of both the organisms and the sampling techniques will help you to answer questions successfully. You may be asked about habitats that you have not studied or you may be required to analyse unfamiliar data, so the more thorough your understanding, the better prepared you will be.

A more general review of Unit 3, particularly Topics 3.2 (Ecosystems), 3.3 (Energy flow) and 3.6 (Human influences on the environment) would be extremely useful before starting your learning of the rest of the topic.

As so much of this topic is common to both specifications, you are recommended to try the relevant exercises and practice questions in the chapter Unit 5B. These are listed at the end of the Checklist and page references given. Suitable assessment questions relating to the Biology (Human) specification have been selected for you to attempt.

Checklist of things to know and understand

Before attempting to answer any of the questions, check that you know and understand the following:

Distribution of plants and animals

❏ the effects of biotic and abiotic factors on the distribution of organisms in a terrestrial and aquatic habitat

❏ appropriate qualitative and quantitative field techniques used in an investigation of the distribution of organisms in a specific habitat

Succession

❏ that ecosystems are dynamic and subject to change over time, illustrated by the change from grassland or abandoned farmland to woodland

❏ the seral stages in a succession

❏ plagio and climatic climax

Control of insect pests

❏ how insect populations can be controlled by biological and chemical methods

❏ the relative advantages and disadvantages of these methods

❏ the bioaccumulation of non-biodegradable toxins

❏ the use of integrated pest management (IPM)

Conservation

❏ the management of grassland and woodland habitats to maintain or increase biodiversity as illustrated by mowing, scrub clearance, use of fire and coppicing

❏ how intensive food production may affect wildlife

❏ how farming practice can enhance biodiversity

❏ the significance of the EU Habitats Directive concerning the conservation of natural habitats and of wild flora and fauna

❏ the significance of Natura 2000.

Practicals

You are required to have carried out practical work to investigate:

❏ *the distribution of plants and animals in at least one habitat*

❏ *the effects of abiotic and biotic factors on them.*

Unit 5H

 Practical work – Helpful hints

You are expected to use sampling methods to study the distribution of plants and animals in at least one habitat. The nature of this investigation will depend on the types of habitats you study.

● Plan your investigation carefully before you start.

● Be aware of safety issues – this is particularly important when working in aquatic habitats.

● Decide whether you are going to use random sampling or systematic sampling.

● Quadrat frames (for example, 0.25 m²) are used to sample the area under investigation.

● When describing vegetation, you may decide to use a subjective estimate of percentage cover or an objective estimate, such as percentage frequency or biomass.

● For sampling in aquatic habitats, techniques such as sweep sampling, or kick sampling are used.

● Terrestrial animals may be collected using a Tullgren funnel or a pitfall trap.

● Environmental measurements in terrestrial habitats include edaphic factors, such as soil depth, moisture content, humus content, pH and mineral content.

● Environmental factors in aquatic habitats include substrate type, temperature, current velocity, light, and dissolved oxygen.

Testing your knowledge and understanding

You can test your knowledge and understanding of this topic by answering the following questions, to be found between pages 57 and 64.

Distribution of plants and animals

Questions 3B.2 and 3B.3

Succession

Questions 3B.4 and 3B.5

Control of insect pests

Question 3B.7

Conservation

Question 3B.8 to 3B.13 inclusive.

Practice questions

Questions 1, 2 and 3.

Unit 5H

⇨ Mark allocations are given for each part of the questions and the answers are given on pages 128–9.

Assessment questions

1   Fabry's disease is a sex-linked recessive genetic disorder that causes mental retardation.

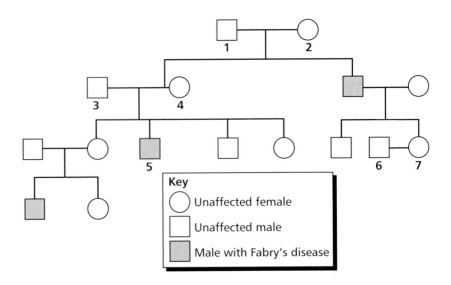

A study was carried out into the inheritance of this disorder in a family, and the results are shown in the pedigree below.

(a) Using the symbol **A** for the dominant allele and **a** for the recessive allele, state the genotype of person 2 **[1]**

(b) Using the evidence from the pedigree, explain why Fabry's disease is described as a sex-linked recessive genetic disorder. **[3]**

(c) Explain why person 3 is unaffected but why one of his children (person 5) has Fabry's disease. **[3]**

(d) What are the chances of the children of persons 6 and 7 having Fabry's disease? Give reasons for your answer. **[4]**

(e) Describe **one** way in which inherited disorders, such as Down's syndrome, may be detected in the developing fetus. **[2]**

(Total 13 marks)
(Modified Edexcel 6041, B/HB1, January 2001, Q.6)

2   The diagrams below show the skeletons of the feet of a modern human, a gorilla and the known parts of a fossil Australopithecine.

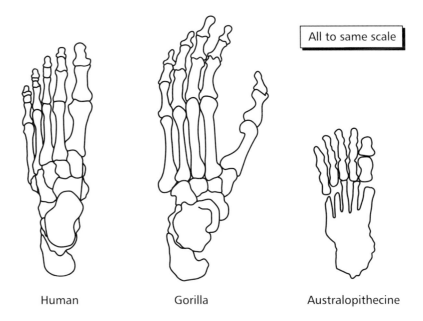

All to same scale

Human   Gorilla   Australopithecine

(a) Explain how the differences between the feet of gorillas and humans are related to their locomotion. **[3]**

(b) What can you suggest about the locomotion of Australopithecines from the structure of the foot? Give reasons for your answer. **[3]**

(Total 6 marks)
(Edexcel 6049, HB2, June 2001, Q.5)

Unit 5H

3   Read through the following passage about the cultural development of stone tools, then write on the dotted lines the most appropriate word or words to complete the passage.

The use of stone tools dates back to million years to *Homo habilis*, who used simple pebble tools. Stone tools are made by hammering one stone against another to knock off flakes and create a sharp edge. This is called .... The first known pebble tools belong to the Oldowan culture.

The tools belonging to the Acheulian tool culture, dating back 1.5 million years, consist of core stones from which flakes had been removed on two sides. These tools were used by *Homo* ....

The tools used by .... man showed further refinement. The tools were made from flakes struck from a core stone. The flakes had razor sharp edges and were ideal for .... These tools belong to the Mousterian flake culture.

More recent tools from the Upper .... , as used by Cro-Magnon man, consisted of long flakes struck from a cylindrical core stone. These flakes were reworked to form knives and saws.

(Total 5 marks)

(Edexcel 6049, HB2, January 2001, Q.1)

6 The GCE Advanced Assessment

The Advanced (A2) half of the specification builds on and adds to the knowledge you gained from studying the Advanced Subsidiary (AS) units. It introduces some of the more difficult concepts, such as metabolic pathways, and you have to show that you have more advanced analytical skills and levels of understanding. In this specification, the Advanced GCE also provides you with an opportunity to select and study an optional topic, investigating some of the applications of Biology to everyday life. The choice of the option is discussed in the Introduction to Unit 4 (page 00).

You also have to show that you can bring together principles and knowledge from all the different areas of the subject, from your AS studies as well as from the A2. This is known as **synoptic assessment** and it is examined in the Unit 5 and Unit 6 written tests. The main thing is to realise that you need to keep your AS notes and keep referring back as you tackle the topics in the A2. It will not be too difficult because many of the topics are related or you will find that you need to use the information from the AS in order to be able to understand the A2. A good example to illustrate this is the topic of **cellular respiration** in Unit 4: you need to know about carbohydrates, enzymes and mitochondria from Unit 1 in order to understand how organisms generate ATP for cell activities. It is also helpful to appreciate how molecules and ions are transported around cells and how gases are exchanged with the external environment from Unit 2.

The unit tests for Unit 4 and Unit 5 are similar to those for the AS units, in that the specification content for those units is examined. As with the AS, the questions are a mixture of those that test your knowledge and understanding and those that require you to analyse and interpret data. Some of the questions in the Unit 5 test will contribute to the synoptic assessment, so you should be prepared for these. As already mentioned, they may require you to draw on your knowledge from any part of the specification, with the exception of your option.

There is no separate specification for Unit 6, which consists of two assessment components as follows:

- a synoptic written paper
- *either* an individual study (T2)
  or a written test (W2) alternative to the individual study.

The synoptic paper

The synoptic written paper, lasting 1 hour and 10 minutes, has two compulsory questions of the longer structured type and an essay. The structured questions test mainly the application of knowledge and biological principles, and the topics could come from any part of the specification. You will definitely need to draw on your knowledge of more than one topic and usually more than one unit. For example, you could find a question on photosynthesis that might refer to material studied in Unit 1 (molecules, enzymes, cell structure), Unit 2 (gas exchange, adaptations to the environment) and Unit 3 (ecosystems), as well as the actual metabolic pathway studied in Unit 5. The questions could include unfamiliar data or situations, which require analysis and which may require you to suggest explanations based on your biological knowledge. You are not expected to know the correct answers to such questions, but you should use your judgement to make a reasonable suggestion. The best way of preparing yourself for this type of question is to practise answering as many questions as possible. There are some at the end of this chapter for you to try and there are others in past papers.

When attempting a synoptic question during your revision, it is a good idea to read through the whole question first and see if you can identify the specification topics on which it is based. This will help you to select the information that you will need in order to write a full answer. Then try answering it. When you think you have answered it, check your answer with the mark scheme. It is particularly important to read through all the points given in the mark scheme because there are often different ways of answering a question. If you have been asked to 'Suggest *one* reason for ....', check that you have given a correct one and also check what other possible answers there were: some of them may not have occurred to you and you could add these to your revision notes. Sometimes the mark schemes do not contain *all* the possible answers, because of space restrictions, but they should give you an indication as to whether you are thinking along the right lines. In very 'open-ended' or general sections of questions, candidates have been known to come up with answers that the Examiners may not have considered. Provided that there is a sound, relevant biological basis to these answers, then credit could be gained. It is always worth attempting an answer to such sections of questions, rather than leaving blank spaces in you answer booklet. The correct strategy is to answer all the parts of the question that you can and then come back to the more difficult sections, making as sensible an attempt as you can.

The essay titles on the synoptic paper will be quite broad and also require reference to material from more than one specification unit. Essays differ from the free-prose questions set on the other written tests and the skills involved are considered to be part of the synoptic assessment. You need to demonstrate that you can:

- describe and explain biological systems and processes
- show an understanding of biological principles and concepts
- sustain an argument
- present evidence for and against a statement
- show an awareness of the implications and applications of modern biology.

You may not be expected to do all these in one essay – it will depend on the title.

You can also gain credit for:

- the selection of relevant material
- the development of a coherent argument
- the quality of your written communication.

> This advice could apply to all parts of questions which ask 'Suggest reasons (or explanations) for ....', not just on the synoptic paper. Remember that synoptic questions are only set on topics and material that is common to the Biology or Biology (Human) specifications and so they will not include any reference to topics in your chosen option.

How can I get a good mark for my essay?

As the synoptic paper for this specification only has three questions, the essay marks form a greater proportion of the total mark. You will have a choice of two essay titles, so you need to spend a little time choosing the right one. If necessary try jotting down a few ideas for each title until you are sure which one you are going to answer. It is also helpful at this point to look at the mark allocation and to consider how the marks are awarded. All the other questions on the written tests are marked according to a points scheme: you need to write down a specific piece of information or carry out a specific task for the award of one mark. Essays are not marked to a points mark scheme, but marks are awarded for the way in which different areas of the topic are presented and discussed (scientific content), the breadth of treatment of the topic (balance) and your written style (coherence). You can gain a maximum of 13 marks for scientific content, 2 marks for balance and 2 marks for coherence. The overall maximum for the essay question is 15 marks, so you can see that the way in which you present your answer and the correct balance of information can enable you to gain a high mark even if you do not include every bit of information about the subject. Reference to the example given later in the chapter will show you how this works.

The scientific content of your essay is important because this is where you can gain most of your marks, so you should spend some time planning your answer. Jot down some paragraph headings and draw a spider diagram or a table of comparisons before you consider the logical sequence of what you are going to write. Remember that it is the quality of your answer which is important and not the number of pages you cover. Do not spend too much time on your plan, or make it too detailed and complicated. You might run out of time and not get full credit for the information you know if you fail to complete your answer. Examiners do read the plans for unfinished essays, but your ideas will need to be very clearly set out in order for you to gain much credit in this way. It is far better to make a simple plan, stick to it and complete your answer.

When you are planning your essay, you should think about how you are going to present your material. Your essay needs an introductory paragraph, setting the context of the topic and making some general points. This should be followed by the development of the topic, written in paragraphs, checking that you do not repeat yourself and that your information is relevant. At the end, it is customary to finish with a concluding paragraph that sums up the major points you have made. It is worth noting that many synoptic essay titles contain references to 'living organisms' or 'in plants and animals', so you should bear this in mind when planning your answer.

There is a great temptation to include diagrams in essays, but too often they either only repeat what has been written or are irrelevant. A good diagram should add to the account or help in the illustration of a point you are making. One example of where a diagram is helpful is in the explanation of a predator-prey relationship, where, with a simple graph, you can indicate the nature of the predator and the nature of the prey and also show how the population numbers fluctuate over time. This could take a long time to explain clearly in prose, but can be shown easily on an annotated diagram. On the other hand, complex diagrams of, for example, heart and kidney structure, are time-consuming and rarely relevant in the context of most essay titles. Bear in mind that diagrams can be of great value in free-prose answers, particularly if they are well annotated.

One further point, while considering your style and presentation – do make your handwriting legible. If the examiner cannot read what you have written, then you cannot be given any credit. Of course, this applies generally to any written answers, but it is particularly relevant to the essay, where you are likely to write at length. If you are aware that your writing is difficult to read – and your teacher or tutors may have spoken to you about this already – then do your best to improve it before the written tests. Nowadays, many students use word-processors for much of their coursework and it is easy to get out of the habit of writing legibly and clearly. Choose a pen you are happy with and make sure you have spare ink cartridges or a new ballpoint pen with you in the examination. Fading ballpoint pen or pencil can be quite difficult to read; remember that most examiners are going to be doing their marking in the evening, when the light is fading.

There is no set specification for the synoptic paper, so it is impossible to draw up a checklist of things to know and understand, as we have done for previous units. Before you attempt the following questions, it would be worthwhile checking that you:

- know the common material of Units 4 and 5 thoroughly
- have reviewed the topics in the AS Units 1, 2 and 3
- have read through your practical work.

Practice questions

Mark allocations are given for each part of the questions and the answers are given on pages 122–3.

The following questions have been chosen and devised to give you an indication of the nature of synoptic testing. You might like to decide for yourself which units are covered in each question.

1   The table below shows the average concentrations of four substances, urea, glucose, sodium ions and potassium ions, in different regions of a kidney tubule (nephron). The figures are expressed in g dm^{-3}.

Substance	Concentration of substance / g dm^{-3}			
	Bowman's capsule	Proximal convoluted tubule	Distal convoluted tubule	Collecting duct
Urea	0.3	0.55	6	15
Glucose	0.1	0	0	0
Sodium ions	0.33	0.33	0.1	0.33
Potassium ions	0.17	0.02	0.06	0.85

(a) Large volumes of water are reabsorbed into the blood from the proximal convoluted tubule. Suggest why the concentration of sodium ions remains unchanged but the concentration of urea increases. [2]

(b) Suggest reasons for the changes in the concentrations of the sodium ions and the potassium ions in the distal convoluted tubule and the collecting duct. [3]

(c) The concentration of glucose falls to 0 in the proximal convoluted tubule. Explain how this is brought about. [2]

(d) The diagram below shows a section through a cell in the wall of the proximal convoluted tubule of a kidney nephron.

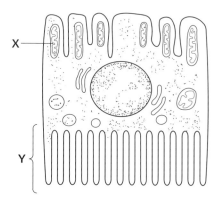

Name the features labelled X and Y on the diagram and indicate their role in the process you described in (c). **[4]**

(e) Name **two** structures, other than those labelled X and Y, which indicate that this is a eukaryotic cell. **[2]**

(Total 13 marks)
(Edexcel B6, January 1998, Q.5)

2   The diagrams below show cell A with two pairs of chromosomes and cells B, C and D which have arisen from cell A by cell division.

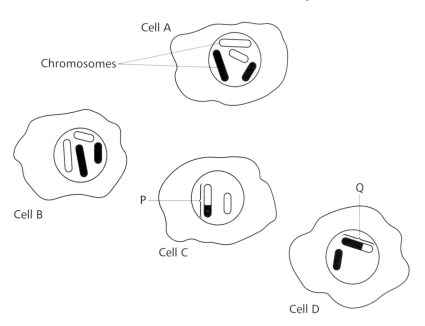

(a) For each of the cells B and C, identify the type of cell division which has occurred to produce the cell. In each case give a reason for your answer. **[2]**

(b) Explain the reasons for the differences between the parts labelled P and Q in cells C and D. **[3]**

(c) Warfarin is a pesticide which is used to kill rodents such as rats. Some rodents are resistant to warfarin and this resistance has been shown to be controlled by a gene with two alleles, **W¹** and **W²**.

Unit 6

The table below shows the possible genotypes of the rats and whether or not they are resistant to warfarin.

Genotype	Phenotype
W^1W^1	Susceptible to warfarin
W^1W^2	Resistant to warfarin
W^2W^2	Resistant to warfarin

In this type of question there is much to read. Think carefully about the first part which is relatively straightforward and deals with material from Units 1 and 2. Then read the part about the warfarin through before answering the questions. With unfamiliar material such as this, some of the answers to the questions will draw on information given in the introduction.

Determine whether the allele for resistance to warfarin, W^2, is dominant or recessive, giving a reason for your answer. **[2]**

(d) Rats with the genotype W^2W^2 require much more vitamin K in their diet than those with the other genotypes.

In a study of the rats around Welshpool, where warfarin had been used, 74 rats were trapped: 42 of them were found to have the genotype W^1W^2, 28 had the genotype W^1W^1 and only 4 had the genotype W^2W^2.

Give a reason why each of the homozygous genotypes is at a disadvantage compared with the heterozygous genotype. **[2]**

(e) A further study was carried out in a nearby area where warfarin was also used. In 1973, nearly 60% of the rats were resistant to warfarin. The use of warfarin was then discontinued and after two years the number of resistant rats had dropped to less than 40%. Explain why the number of resistant rats dropped after the warfarin was discontinued. **[3]**

(f) Explain why the allele for resistance to warfarin is likely to remain in the gene pool when the use of warfarin is discontinued. **[2]**

(Total 14 marks)
(Edexcel B6, June 1996, Q.2)

The individual study (T2)

For most students, the second part of Unit 6 consists of an individual study. This piece of practical work gives you the opportunity to plan and carry out a scientific investigation on your own. You need to write a report on the investigation and it is recommended that you adopt a style similar to that used in scientific journals. The study should be based on a quantitative investigation or experiment and can be linked to any part of the specification – it does not have to be based on material in Units 4 or 5. It should be stressed that the investigation must be *individual*, based on a topic that either interests you or has developed from another experiment you might have done as part of your course. It is not expected that straightforward experiments, such as those listed in the specification, are suitable, but some development or extension of such an experiment could form an interesting study. The practical work associated with the AS (T1) requires you to work individually, but your investigation could be similar to others. In T2, you must work on your own and there should be no sharing of results.

Another difference between T1 and T2 is that you need to use a simple statistical test in the analysis of the data you obtain. You have to bear this in mind when planning your investigation, so that you can take a suitable number of samples or readings. A well-planned and executed individual study can contribute much to the overall grade you achieve, so it is worthwhile putting some effort into choosing a suitable topic and carrying out the investigation carefully. Your teacher can advise you on possible topics and further information about experimental techniques, statistical tests and advice on planning, using data and writing up the report is given in *Tools, Techniques and Assessment in Biology*, the Nelson Advanced Science course guide for Biology.

Your teacher assesses your coursework for T1 and samples are inspected and moderated by the awarding body, Edexcel. For T2, some assessment is done by your teacher, but the complete written report must be submitted to Edexcel for marking. The criteria for the award of marks are very similar to those for T1, but there is greater emphasis on the analysis of evidence, including the presentation of results and their statistical treatment, the discussion and the evaluation. You also gain credit for the account of the methods used, the inclusion of background information and the style of your report. The reports are not expected to be excessively long and a limit of about 3500 words has been suggested. You will not be penalised if your report is longer than this, but you should appreciate that you do not need to spend excessive amounts of time either on the practical work or the writing up in order to gain a good mark. Many of the best studies start with a simple, straightforward hypothesis, investigated carefully using the resources available and yielding meaningful results to which an appropriate statistical test can be applied to assess their significance.

Written alternative to the individual study (W2)

If you are unable to carry out an Individual Study or your centre chooses not to enter for T2, it is possible to take a written test (W2). This paper contains questions based on practical skills, data handling and the interpretation of results. You will be required to demonstrate that you are familiar with experimental techniques because the questions set test your ability in designing investigations, using statistical tests and analysing data.

There will be two questions on this paper. One question could require you to organise some data into a table, then to present that data in an appropriate graphical form. You may be asked to suggest a suitable statistical test and to indicate what conclusions could be drawn. It is likely that the data will be unfamiliar to you, but you should be able to organise it using your own experiences of practical work.

The second question will suggest a hypothesis and ask you to plan an investigation to test that hypothesis. Details of planning are expected, together with some idea of what results would be expected and how you would present and analyse them. Finally, you could make suggestions about the limitations of the method you have chosen and any further work that you consider could be undertaken. The details expected are similar to the criteria set out for the individual study. You need to be familiar with practical techniques and the decision to attempt this paper should not be a substitute for carrying out practical work.

Reference to past W2 papers, to the specimen papers and to the practice questions given below will give you an indication of the types of questions set on W2. The mark schemes show a range of acceptable answers and the knowledge required to achieve success. Although this paper is an alternative to the individual study, it is by no means the easy option, and you do not have the satisfaction of carrying out your own investigation.

 Practice questions

> Mark allocations are given for each part of the questions and the answers are given on page 123.

1 A group of students carried out an ecological investigation into the distribution of two species of trees in a wood. They found that one species (A) was more common on dry, well-drained soils whilst the other species (B) was more common where the soil was wet and poorly drained. They produced the hypothesis that one reason for this was that the leaves of species B lost water vapour more quickly than the leaves of species A.

To test this hypothesis they collected a sample of 100 g of leaves from species A and a sample of 100 g of leaves from species B. They then hung each sample on a line to dry in identical conditions in the laboratory. Both samples were then reweighed each hour for five hours.

Species A						
Mass after	1h	2h	3h	4h	5h	
	92.5	81.0	64.0	57.2	51.6	in g
Species B						
Mass after	1h	2h	3h	4h	5h	
	72.5	45.9	31.0	24.1	20.8	in g

An extract from the records of this investigation is shown below.
(a) Calculate the loss in mass compared to the original mass, for each sample every hour. Then organise the data in a suitable table so that the loss in mass for each sample can be compared. **[4]**

(b) Use the data in your table to present this information in suitable graphical form. **[4]**

(c) What conclusions can you draw from the results of this investigation? **[2]**

(Total 10 marks)
(Edexcel WTA2, June 1998, Q.2)

2   Aphids are insect pests of glasshouse crops, such as green peppers (*Capsicum* sp.).

Plan an investigation, which you personally could carry out, to test the hypothesis that the presence of aphids significantly reduces the yield of green peppers grown in a glasshouse.

You should give your answer under the following headings:

(a) Plan of the investigation to be carried out   [9]

(b) Recording of raw data measurements, presentation of results and methods of data analysis   [7]

(c) Limitations of your method, and an indication of further work which could be undertaken   [5]

(Total 21 marks)
(Edexcel WTA2, June 1998, Q.3)

Answers

Unit 4 Respiration and coordination and options

Core section: Respiration and coordination

Topic 1 Metabolic pathways

Testing your knowledge and understanding

1.1 *metabolic pathway:* a sequence of chemical reactions in a cell, each reaction is catalysed (or controlled) by a specific enzyme;

1.2 oxidoreductases – enzymes which catalyse oxidation and reduction reactions, such as removing hydrogen from a substrate (oxidising the substrate); hydrolases – enzymes which catalyse hydrolysis reactions;

1.3 *anabolism:* the synthesis of complex molecules from simpler ones, such as protein synthesis; *catabolism:* the breakdown of complex molecules into simpler ones, such as cellular respiration;

1.4 providing energy for anabolism, the light-independent reaction of photosynthesis, active transport, muscle contraction, and cell division; (you could also include bioluminescence – the light from a glow-worm – a species of beetle, not a worm at all – originates from the breakdown of ATP)

1.5 glycolysis, the Krebs cycle and oxidative phosphorylation;

1.6 you should start with glucose, then show that this is converted to glucose 6-phosphate; This is converted to glycerate 3-phosphate, then to pyruvate;

1.7 pyruvate, ATP and reduced coenzyme (NADH + H⁺);

1.8 in the cytoplasm;

1.9 you should show that oxaloacetate (a 4-C compound) combines with acetyl CoA to form citrate (a 6-C compound); citrate is then converted back into oxaloacetate, forming a cycle of reactions;

1.10 carbon dioxide, reduced coenzymes and ATP;

1.11 in the matrix of a mitochondrion;

1.12 ATP is formed as electrons are passed via a series of carriers to oxygen; during this process, the flow of protons back into the mitochondrial matrix provides the energy for the synthesis of ATP;

1.13 they combine with oxygen to form water;

1.14 ethanol, carbon dioxide and ATP;

1.15 lactate and ATP (no carbon dioxide is produced);

1.16 two molecules of ATP per molecule of glucose;

1.17 you should show the *outer membrane*, the *inner membrane* folded to form *cristae*, the *intermembranal space* and the *matrix*; the Krebs cycle takes place in the matrix; oxidative phosphorylation in the inner membrane;

Practice questions

1 (a) ATP; [1]

(b) mitochondrion; [1]

(c) carbon dioxide; ATP; reduced coenzymes; [2]

(d) pyruvate is first converted to ethanal; carbon dioxide is produced; then ethanal is converted to ethanol; [3]

Examiner's comments

In Question 1(a), the abbreviation ATP is acceptable. In (b), stage 4 represents the conversion of pyruvate to acetyl coenzyme A (the link reaction), which occurs in the mitochondrial matrix, but mitochondrion is an acceptable answer. There are several acceptable answers to (c); you could name one or more of the reduced coenzymes, you have to name two different products to gain both marks. To answer (d) fully, think through the sequence of events in anaerobic respiration in yeast, remembering that the final products are carbon dioxide and ethanol, but ethanal is formed first. Be careful not to confuse *ethanal* and *ethanol*! Ethanal is also known as acetaldehyde.

(Total 7 marks)

2 (a) (i) carbon dioxide; [1]

(ii) ethanol; [1]

(b) the hydrogen atoms are picked up by a coenzyme / by NAD; then used to reduce ethanal to form ethanol; [2]

(c) ATP; lactate; heat; [2]

Examiner's comments

In Question 2(a)(i), you could have given the chemical formula for carbon dioxide. In (a)(ii), identifying ethanol as a product of anaerobic respiration in yeast is fairly straightforward; ethanol is also known as ethyl alcohol, which is an acceptable answer. In (b), remember that this is about *anaerobic* respiration, the hydrogens are picked up by a coenzyme (NAD) and then reduce ethanal to form ethanol. This process oxidises the reduced coenzyme back to NAD+. In (c), 'heat' is an acceptable answer – muscles do get hot during vigorous activity.

(Total 6 marks)

3 (a) (i) they are picked up by a coenzyme / NAD; then reduce pyruvate to form compound X; **[2]**

(ii) lactate; **[1]**

(b) to oxidise the reduced coenzymes; so that glycolysis can continue / so that more ATP can be formed; **[2]**

Examiner's comments

Question 3(a)(i) is straightforward, but remember that this question is about *anaerobic* respiration, so references to oxidative phosphorylation are inappropriate here. In (a)(ii), lactic acid is an acceptable alternative to lactate. Part (b) is a little harder, but remember that if the reduced coenzymes were not converted back to their oxidised form, glycolysis would eventually stop. Conversion of pyruvate to lactate (or to ethanol in yeast) solves the problem.

(Total 5 marks)

4 (a) (i) they combine with oxygen; to form water; **[2]**

(ii) the inner membrane; **[1]**

(b) glycolysis / the Krebs cycle; **[1]**

(c) phosphorylation of a substrate / formation of glucose 6-phosphate; active transport / ion pumps; muscle contraction; the light-independent reaction; protein synthesis; cell division; synthesis of organelles; **[2]**

Examiner's comments

In Question 4(a)(i), remember that water is the final product of aerobic respiration. In (a)(ii), cristae is acceptable as an alternative to inner membrane. There are several acceptable alternative answers in (b), link reaction is also correct. Again, there are a number of acceptable answers to part (c), you could name macromolecules other than proteins, for example. Remember to give specific uses of ATP, an answer such as 'ATP is used as energy' is really too vague.

(Total 6 marks)

Topic 2 Regulation of the internal environment

Testing your knowledge and understanding

2.1 *homeostasis:* the maintenance of a constant internal environment;

2.2 osmoregulation and excretion of nitrogenous excretion;

2.3 your diagram should include *Bowman's capsule* and *glomerulus, the proximal convoluted tubule, loop of Henlé* (with the descending and ascending limbs correctly labelled), the *distal convoluted tubule* and *collecting duct*;

2.4 proteins, because protein molecules are generally too large to pass through the glomerular filter;

2.5 glucose;

2.6 you could name any two of: water, amino acids, urea and mineral ions;

2.7 enables the kidney to produce a high solute concentration in the medulla; this allows water to be drawn out of the collecting ducts, producing a concentrated urine;

2.8 they have exceptionally long loops of Henlé; this increases the multiplier effect, enabling the mammals to reabsorb water very effectively and produce a small volume of very concentrated urine;

2.9 high – to increase the reabsorption of water in the collecting ducts; this helps to conserve water when, on a hot day, more will be lost by sweating;

2.10 the urea concentration would increase, because the excess amino acids, from the digestion of protein, will be deaminated in the liver; this leads to increased synthesis of urea;

2.11 breakfast usually contains carbohydrates (think of the starch in toast or breakfast cereals!); after digestion of carbohydrates, the glucose which is formed is absorbed into the blood stream;

2.12 the increase stimulates the β-cells in the islets of Langerhans to secrete insulin, which lowers blood glucose concentration;

2.13 glucagon – increases blood glucose; adrenaline – increases blood glucose;

2.14 phytochrome, rhodopsin and iodopsin;

2.15 you could include references to their speeds of effects, transport, and general or specific effects;

2.16 peptide hormones – attach to a receptor on the cell surface membrane, and activate adenyl cyclase, which converts ATP into cAMP; this then activates an enzyme system;
steroid hormones – pass through the cell membrane and form a complex with a receptor molecule; this passes into the nucleus and increases transcription, resulting in increased protein synthesis;

2.17 you should show the *cell body*, containing a *nucleus*, *dendrites*, an *axon* with *myelin sheath*, and *nerve terminals* at the end of the axon. You may also have labelled the *nodes of Ranvier*, the small gaps between adjacent Schwann cells;

2.18 *resting potential:* the overall negative electrical potential of about –70 mV inside the axon, in the resting state; *depolarisation:* occurs when sodium ions rush into the axon, changing the potential from -70 mV to about +40 mV, the action potential;

2.19 you should show the *presynaptic terminal*, containing *synaptic vesicles*, the synaptic cleft, and the *postsynaptic cell membrane*;

2.20 calcium ions enter the presynaptic terminal, synaptic vesicles fuse with the presynaptic membrane, the transmitter substance is released by exocytosis, the transmitter substance diffuses across the synaptic cleft and interact with receptors on the postsynaptic membrane;

2.21 acetylcholine (ACh), adrenaline and noradrenaline are the most likely ones. There are many other transmitter substances, including dopamine, 5-hydroxytryptamine (5-HT or serotonin) and gamma aminobutyric acid (GABA);

2.22 check your answer using a textbook as there are many acceptable functions you could include;

2.23 *spinal reflex:* a rapid, automatic response to a stimulus, involving nerves which pass through the spinal cord, without conscious control of the brain. An example of a spinal reflex is the familiar *knee jerk reflex*]

Practice questions

1 glomerulus; glucose / water / urea / amino acids; ultrafiltration; glucose; water;

(Total 6 marks)

Examiner's comments

This question requires specific knowledge about kidney function. Notice that the second sentence refers to *molecules*, so examples of ions would be incorrect here. It does not matter whether you write glucose, or water, first. There are a number of other small molecules which would also be acceptable, such as uric acid, ammonia, vitamin C and creatinine.

2 (i) 99.2 (or 99.17); (ii) 0.0 g; (iii) 28 g; (iv) 52.8 (or 52.83);

Examiner's comments

When calculating answers, be careful to avoid rounding up (or rounding down) errors. As a general rule, you should give as many decimal places in your answer as there are in the figures in the question, so in part (i), for example, '99.2' is acceptable, but not '99'. It is interesting to note that about half of the urea which is filtered out is reabsorbed; urea is an important solute which helps to maintain the osmotic gradient for reabsorption of water in the medulla of the kidney.

(Total 4 marks)

3 low; rhodopsin; opsin; dark;

(Total 4 marks)

Examiner's comments

This question requires specific knowledge of the function of rod cells. Rhodopsin is also sometimes known as visual purple so this would be an acceptable alternative in the second space.

4 (a) *action potential:* a brief depolarisation; of the axon membrane; from about –70 mV to about +40 mV; due to the influx of sodium ions / increased permeability to sodium ions; [2]

(b) *transmitter substance:* a chemical / named example released by the presynaptic neurone; into the synaptic cleft / diffuses across the synapse; binds to a receptor on the postsynaptic membrane / causes a change in the permeability of the postsynaptic membrane; [2]

(c) *myelination:* produced by Schwann cells; which wrap around the axon; producing layers of cell membranes; myelination insulates the axon / speeds up conduction of the impulse; [2]

(Total 6 marks)

Examiner's comments

Note that in each of the above answers, there are three or four points for a maximum of 2 marks. It is preferable to give a full explanation of each term, to be sure of picking up full marks.

5 (a) A (left) = cerebral hemisphere / cerebrum; B = medulla oblongata; C = cerebellum; [3]

(b) (Any *two* of the following) control of: heart rate; breathing; peristalsis; vasoconstriction / vasodilation; blood pressure; swallowing; salivation; sneezing; coughing; vomiting; [2]

(Total 5 marks)

Examiner's comments

The two cerebral hemispheres make up the cerebrum, and either name is acceptable here. Be careful not to confuse *cerebrum* with *cerebellum*. Any two functions of the medulla oblongata are acceptable in part (b), but remember to state '*control of*' before naming the function.

6 (a) 30 μmol dm^{-3}; [1]

(b) (i) the blood glucose concentration increases; rapidly / from 5.0 to 7.2 mmol dm^{-3}; because glucose is absorbed from the gut / glucose enters the blood stream; [2]

(ii) the blood glucose concentration decreases; steadily / from 7.2 to 4.2 mmol dm^{-3}; because glucose is taken up by cells / liver / muscle / converted to glycogen for storage; [2]

(c) they both increase up to 30 minutes / both highest at 30 minutes; then they both decrease; the rise in blood glucose concentration stimulates the release of insulin; from the β-cells in the islets of Langerhans; [3]

(d) hormone – glucagon; effect – increases blood glucose concentration; **[2]**

(Total 10 marks)

Examiner's comments

In answers to questions which involve reading values from a graph, or calculating a value, you should always give appropriate units. In part (a) of this question, be careful to use the correct axis for the concentration of insulin. Part (b) requires descriptions and explanations for the changes. When you are interpreting a graph, it is always a good idea to quote figures read from the graph, to support your description. To answer part (c) correctly, you need to think about the sequence of events as glucose is absorbed into the blood stream; the rise in glucose concentration stimulates the β-cells in the islets of Langerhans to secrete insulin, which accounts for the corresponding rise in the blood insulin concentration. There are other acceptable answers to part (d), for example, you might have named glucocorticoids, or adrenaline, both of which increase blood glucose concentration. Be careful with the spelling of *glucagon* so that the word is not confused with *glycogen*!

Option A Microbiology and biotechnology

Testing your knowledge and understanding

A1.1 below are given suggested headings which should help you to organise your own revision notes. Make sure you tease out and learn the details as well as build up the overall framework.

size – find out a range of sizes for bacteria, fungi and viruses; is the term 'microorganism' justified for all under this heading?

acellular / multicellular – how far does each group fit under these headings?

eukaryote / prokaryote – which does each group fit under (or not)?

subcellular structure – compare bacteria with a typical plant cell then with fungi and viruses as far as you can (use your annotated diagram to help you); give details of organelles, cell wall structure and any other features;

nutrition – requirements considered in some detail in 'Culture techniques'. Link this also to activities of bacteria that are of importance to the environment and exploited in biotechnology.

reproduction – refer to methods of reproduction, whether asexual or sexual and the special methods used by viruses to replicate. Link this also to the spread of those bacteria and viruses which cause disease and to consideration of growth of populations

of microorganisms in industrial fermentations for products to be harvested (biotechnology!)

A2.1 (a) check details of how you would implement 'sterile' techniques in the laboratory and arrange in sensible order in flow chart – refer to clean area (use disinfectant) to work in, using sterile petri dishes or other glassware (such as pipettes), autoclaving or using pressure cooker for sterilising medium, pouring agar plate and holding lid over it, flame loop and neck of tube with culture, etc. grow inoculated plates for suitable time at required temperature in incubator;

(b) add details of sterile entry of medium and inoculum; system for temperature control (e.g. water jacket for cooling); monitor and control pH by adding acid or base when required; sterile air entry (if aerobic conditions needed); motor to stir / agitate contents; system for collection of product when ready to harvest; etc;

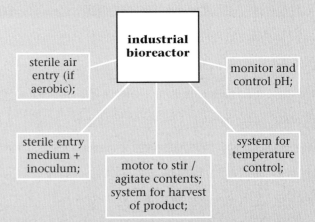

(c) try to work this out for yourself, then see the answers given for mycoprotein and penicillin production in Topic A3.

A3.1 (a) A1 = *malting*; grains; steeped; germinate; grains; malting floor; gibberellic acid; aleurone; 65 °C; *examples of enzyme reactions*: various proteins (e.g. hordein) degraded by proteases to simpler peptides and some to amino acids; carbohydrates (mainly starch) degraded by α- and β-amylases to dextrins and maltose; lipids degraded by lipases to fatty acids; polymers of pentoses in the cell walls degraded by enzymes such as β-glucanase to simpler pentose sugars;

A2 = *milling*; malted grains; starchy; gritty;

A3 = *mashing*; hot water; mash tun; sugars; wort; *examples of enzyme reactions*: amylose and amylodextrin degraded by α- and β-amylases to dextrins and maltose;

A4 = *fermentation*; wort; *Saccharomyces cerevisiae*; *Saccharomyces carlsbergensis*; *comments*: during initial stage of fermentation yeasts respire aerobically, then when oxygen used up respiration is anaerobic, producing ethanol;

(b) helps provide air to give oxygen for respiration and remove carbon dioxide;

(c) when boiling wort with hops in the copper (see below);

(d) after mashing, when boiling wort with hops in the copper; boiling sterilises the liquid, stops enzyme activity, extracts tannins and oils from hops which contribute to flavour; copper;

(e) *ales:* 20 °C, 5 days; *lagers:* 5 to 15 °C, between 7 and 14 days;

(f) conditioning / maturation; artificially carbonated; bottled / put in casks etc; sometimes pasteurised;

(g) spent grain (from mash tun) for cattle feed; spent hops (from copper) for fertiliser; surplus yeast (from fermenting vessel) for other food products (e.g. yeast extract);

Option B Food science

Testing your knowledge and understanding

B1.1 **(a)** *the 'feature' in fast food is given in italics, followed by reasons why 'it might be worrisome'*

high calories – means high energy value – suggests high carbohydrate and / or fat intake, excess converted to fat and stored; excess fat increases body weight, leads to overweight / obesity etc.;

high sodium – high sodium intake associated with high blood pressure;

high fat – excess is stored – leads to overweight / obesity etc.;

high cholesterol – high cholesterol concentration in plasma may increase risk of atherosclerosis (accumulation of lipids etc. on walls of arteries, thus narrowing lumen) hence increased risk of CHD;

excessive consumption (more than recommended daily intake) – if high carbohydrate and / or fat intake, excess converted to fat and stored; excess fat leads to overweight / obesity etc.;

high saturated fats – likely to lead to increased synthesis of cholesterol (see above);

low fruit and vegetables – these contain non-starch polysaccharides (NSPs) – important (to include / eat fruit and vegetables) to decrease 'transit time' of food and reduce risk of bowel diseases; may also reduce blood cholesterol (see above);

(b) the 'conditions' listed are obesity, coronary heart disease, hypertension, diabetes, cancer – most of these (except diabetes and cancer) have already been covered in the statements above; increased risk of diabetes (mature onset diabetes mellitus) linked with obesity; increased risk of cancer of colon associated with low NSP etc. (see above);

(c) lack of physical exercise ('sedentary lifestyle' – as

in passage); metabolic factors; genetic factors; psychological factors (e.g. response to stress); etc.; increased risk – see above list + osteoarthritis, infertility (women), difficulties in childbirth, certain cancers;

(d) combination of reducing energy intake and increasing energy expenditure; use of certain 'restricted' diets but care needed in administration of these – avoid drastic reduction + caution against lack of important nutrients; leading to protein, mineral and vitamin deficiencies etc.; certain low calorie diets may give unpleasant side-effects (diarrhoea, nausea, etc.);

B1.2 A = bulimia; eating; self-induced vomiting; weight; fatigue, constipation, upset of electrolyte balance (associated with vomiting), dental erosion, sore throat etc.;

B = scurvy; C; ascorbic; synthesis; wounds; bleeding gums, haemorrhages under the skin;

C = anaemia; iron; haemoglobin; red; myoglobin; red blood; pale;

D = kwashiorkor; protein; swollen; breast-feeding; enlargement of the liver due to accumulation of fat;

E = anorexia nervosa; eating; little; weight; lowered;

F = marasmus; protein; energy; emaciation / wasting; pale / apathetic / lacking in skin fat etc.;

B2.1 **(a)** lactose (glucose + galactose); sucrose (glucose + fructose);

(b) sucrose used as standard and given a relative sweetness (RS) = 1 (the threshold concentration of 0.01 mol dm^{-3} is taken as the minimum concentration of sucrose which produces a response on the tongue); order of sweetness = aspartame (1), fructose (2), sucrose (3), glucose (4), lactose (5);

(c) clean conditions in laboratory; solutions (suitable concentrations) of the substances (e.g. 4%) for the sugars); taste each solution, rinsing mouth with drinking water in between;

B3.1 **(a)** (1) oxygen; low oxygen prolongs storage life but if too low leads to anaerobic respiration and possible undesirable changes in the fruit (such as accumulation of ethanol); (2) carbon dioxide; increased carbon dioxide generally delays postharvest changes; (3) increases relative humidity (in package); increases likelihood of microbial attack;

(b) (1) and (2) lead to faster depletion of respiratory substrate and / or conversion of stored carbohydrate (e.g. starch); shrivelling of fruit (due to loss of mass); sooner onset of deterioration / senescence / other postharvest changes (colour etc.); (3) increased loss of water; loss of turgor / shrivelling of fruit;

(c) need to consider – permeability in relation to gas balance (oxygen and carbon dioxide), accumulation of ethene (and effect on ripening), reducing loss of water; mechanical properties (e.g. strength in relation to tearing, bursting and sealing); stability in relation to environmental conditions, and during processing and packaging; visibility and attractiveness of package to consumer;

B3.2 (a) uses respiratory substrate (e.g. glucose); storage carbohydrate or other reserves converted to suitable substrate; increased carbon dioxide delays postharvest changes; lower oxygen levels prolong storage life (delay postharvest changes), but if too low may get anaerobic respiration (see B3.1(a) above); CA relies on deliberate monitoring and control of gas mixture at desired composition (particularly oxygen and carbon dioxide); MA holds produce in airtight container and respiration of produce changes atmosphere within the container; importance is that gas balance can be adjusted or modified to prolong storage life; packaging – see answer to B3.1(c) (above); waxing (e.g. lemons and other fruits) used to reduce effects of water loss but can result

in development of anaerobic conditions internally (see above for consequences);

(b) loss of turgor, loss of mass (see B3.1(b) above); suitable packaging, storage container, waxing of individual fruit;

(c) sugars – sucrose, glucose, fructose; acids – citric acid, malic acid, ascorbic acid; generally sugars increase and acids decrease as fruit ripens; leads to sweeter taste (or less sourness) in riper fruit; many volatile substances, present in minute quantities, contribute to flavour;

(d) promotes ripening; provision of adequate ventilation in storage chamber to remove ethene produced by fruits; encourages ripening (due to ethene being given off);

(e) often from green through yellows, oranges to reds; pigments include chlorophyll, carotenes, lycopene (tomatoes) etc.;

B4.1

Modified food	Fermentation reactions	Processing essentials
Sauerkraut	sugars (mainly glucose) to lactic acid (~1.25%), pH 3.4 to 3.6	• cabbage washed and shredded • granular salt added (~2.25%, by weight) • anaerobic conditions • temperature ~ 21 to 24 °C • 1–2 weeks
Soya sauce	• two main stages of fermentation • starch and proteins broken down by amylases and proteolytic enzymes • other microorganisms metabolise these producing range of compounds, including synthesis of vitamins	• soya beans soaked in water, boiled and drained • roasted and crushed wheat added • starter culture added, fermented for about one week on trays • brine added and mash transferred to vat • fermentation for 2–12 months • filtered to recover soya sauce
Wine	• glucose → ethanol	• crush fruit to must • treat with SO_2 (sulphite) or pasteurise • stir – aerobic for a few days • then anaerobic (different times and temperatures for red and white wines)
Yoghurt	• lactose (sugar in milk) to lactic acid • low pH (4.0–4.5) causes protein casein to coagulate to soft gel • some other compounds contribute to flavour	• may blend milk or add dried skimmed milk • milk heated to 88 to 95 °C • homogenise, cool to 42 to 47 °C • add starter culture • incubate 3–6 hours, at 37 to 44 °C • cool and add fruit or flavour if required

Option C Human health and fitness

Testing your knowledge and understanding

C1.1 cardiac muscle found only in the heart; cardiac muscle is myogenic; cardiac muscle cells are joined by intercalated discs; (any 2)

C1.2

```
┌──────────────────────┐   ┌──────────────────────┐
│ Contraction of       │   │ Contraction of external│
│ diaphragm            │   │ intercostal muscles    │
└──────────┬───────────┘   └───────────┬──────────┘
           │                           │
           ▼                           ▼
        ┌─────────────────────────────────┐
        │ Increase in volume of thorax     │
        └─────────────────┬───────────────┘
                          ▼
        ┌─────────────────────────────────┐
        │ Expansion of the lungs           │
        └─────────────────┬───────────────┘
                          ▼
        ┌─────────────────────────────────┐
        │ Decrease in intrapulmonary pressure│
        └─────────────────┬───────────────┘
                          ▼
        ┌─────────────────────────────────┐
        │ Pressure gradient established from│
        │ atmosphere to alveoli            │
        └─────────────────┬───────────────┘
                          ▼
        ┌─────────────────────────────────┐
        │          INSPIRATION             │
        └─────────────────────────────────┘
```

Mechanism of inspiration

C1.3 neurone; acetylcholine; action potential; contraction;

C2.1 HR = Q/SV; = 5040/70 = 72 beats per minute;

C2.2 fully oxidised to CO_2 and water; converted back to glucose in liver (gluconeogenesis);

C2.3 glycogen stores used up and maintained at low level for several days (by eating mainly fats and proteins); then carbohydrates reintroduced in diet; and body increases its stores of glycogen above the normal level; so increasing endurance capacity;

C3.1 age; gender; diet; diabetes; cigarette smoking; high blood cholesterol; genetic factors; (any 3)

C3.2 dust particles in lungs lead to fibrosis; causing thickening of the alveolar walls; and reduced efficiency of gas exchange;

C3.3 surgical removal of tumour; radiotherapy using gamma-rays to kill tumour cells; chemotherapy using anti-cancer drugs;

Unit 5B Genetics, evolution and biodiversity

Topic 1B Photosynthesis

Testing your knowledge and understanding

1B.1 (a) A = cuticle; B = upper epidermis; C = palisade tissue; D = spongy mesophyll tissue; E = lower epidermis; F = lateral veins; G = xylem; H = phloem; I = midrib; J = lamina / leaf blade;

(b) **(i)** C; **(ii)** E; **(iii)** B;
(iv) G; **(v)** J; **(vi)** F / I;
(vii) H; **(viii)** A

1B.2

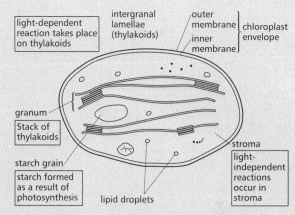

1B.3 (a) light and chlorophyll;

(b) air / atmosphere; diffuses into the leaf through the stomata along a concentration gradient;

(c) non-cyclic photophosphorylation in the light-dependent reaction;

(d) carboxylation;

(e) PGA / phosphoglyceric acid;

Practice questions

1 (a) A = upper epidermis; B = palisade tissue; C = spongy mesophyll tissue; **[3]**

(b) cells elongated / long axis vertical; light trapping reference; many chloroplasts; for photosynthesis / eq.; box-shaped cells / fit closely together; better light capture; thin cell walls; gas exchange / light penetration more effective; cells close to upper surface; light trapping reference; **[4]**

(Total 7 marks)

Examiner's comments

Part (a) of question 1 should be very straightforward. It draws on some knowledge from Unit 1, which is taken further in this unit. Make sure that the functions you give relate to the adaptations of the cells in part (b).

2 (a) X = carbon dioxide; Y = PGA / phosphoglyceric acid; **[2]**

(b) the light-dependent reaction / grana / thylakoid; **[1]**

(c) stroma of the chloroplast; **[1]**

(Total 4 marks)

3

Statement	Light-dependent reaction	Light-independent reaction
Oxygen produced	✓	✗
Carbon dioxide fixed	✗	✓
Occurs in stroma	✗	✓
Uses NADPH + H$^+$	✗	✓
Produces ATP	✓	✗

(Total 5 marks)

4 (a) (i) provides oxygen; for respiration; for active transport; **[2]**

(ii) excludes light; prevents growth of algae; algae would use nutrients; **[2]**

(b) magnesium – chlorophyll formation / cofactor for enzymes; phosphate – phospholipids / ATP / nucleotides / phosphorylation / eq.; **[2]**

(c) reference to at least three different concentrations of nitrate ions in solution; control set of apparatus; same age seedlings used in all; same concentrations of other ions in nutrient solution; same environmental conditions for all; reference to parameter used to judge growth; repetition / replication; **[4]**

(Total 10 marks)

Examiner's comments

This type of question tests your knowledge of the practical work required in the specification. Even if you have not carried out this particular experiment, you should be able to make sensible suggestions from your experience.

---

5 (a) (i) J = 2.5–3.0 %; K = 6 %; **[2]**

(ii) uptake of carbon dioxide increases as light increases; gradient steepest at 5 %; little change after 15 %; correct reference to compensation point; correct reference to relative rate of photosynthesis and respiration; **[3]**

(b) (i) gradient steeper for J; levels off at lower light intensity; maximum rate of uptake is lower / numerical difference given; J reaches compensation point at lower light intensity than K; **[2]**

(ii) J; compensation at lower light intensity / faster rate of photosynthesis at low light intensity; **[2]**

(c) (i) temperature / light quality / wavelength / water availability / air speed / oxygen; **[1]**

(ii) temperature – affects enzyme activity / eq.; light-independent reactions enzyme controlled;

light quality / wavelength – affects rate of photosynthesis / eq.; reference to the wavelengths absorbed by the pigments / eq.;

water – affects stomata / stomata close; so less gas exchange / eq.;

wind / air speed – increased water loss so stomata close; less photosynthesis so less carbon dioxide used;

oxygen – reduces carbon dioxide uptake / increases carbon dioxide release; reference to photorespiration; **[2]**

(Total 12 marks)

Topic 2B Control of growth in plants

Testing your knowledge and understanding

2B.1 *photosynthesis* – synthesis of carbohydrate; increase intensity, increases synthesis (unless other factors become limiting;

phototropism – unilateral light results in bending of shoots towards light (positive phototropism); affects distribution of auxins; position of leaves (oriented to receive maximum light, in relation to direction of sun);

photoperiodism – initiation of flowering; may stimulate germination; detected by phytochrome pigment; interconversion of P_R and P_{FR} dependent on wavelength of light;

2B.2

Plant growth substance	Description numbers
Auxins	*5, 12, 16, 19, 24*
Cytokinins	*2, 13, 14, 17, 23, 26, 32*
Gibberellins	*4, 7, 10, 17, 18, 21, 27*
Abscisic acid	*3, 6, 9, 15, 20, 23, 30, 31*
Ethene	*1, 8, 11, 22, 25, 29*

2B.3 **(a)** promote (usually) – auxins; gibberellins; cytokinins; inhibit (usually) – abscisic acid; ethene;

(b) pick out of the list – see numbers 28 to 32

2B.4 **(a)** stimulation of lateral roots;

(b) linked to apical dominance and effects of auxin;

(c) treatment of unpollinated flowers may lead to development of seedless fruits (auxin produced in pollen linked to stimulation of fruit development after fertilisation);

(d) can delay abscission of fruit or 'fruit drop;

(e) synthetic compounds (e.g. 2,4-D) used as selective herbicides – can alter growth of certain plants (e.g. monocotyledons survive, dicotyledons show distorted growth);

Practice questions

1 **(a)** in C, higher auxin concentration in agar block away from the light than in D or

in C, lower auxin concentration in block close to light than in D;

in D, concentration in both blocks (nearly) the same, in C greater concentration in block away from light than near light;

quantitative comparison between C and D; in C, auxin has moved away from light;

glass prevents movement (of auxin); **[3]**

(b) no;

in D just as much auxin on both sides / eq.;

if hypothesis is correct would expect less auxin on light side than on dark side;

in A and B, almost same / 100 %, auxin collected in light and dark;

in C and D both totals = 100 % / none destroyed; **[3]**

(c) as weedkiller / herbicide;
to kill broad-leaved weeds / dicots / selective;

applied to cuttings;
to promote root development;

to promote fruit growth / produce fruits without fertilisation;
in cucumbers / tomatoes / aubergines / grapes;

increase, flowering / fruit set;
in tomatoes / pineapples;

cause fruit drop;
by spraying early / to thin fruit crops / in olive / apple / pear;

prevent fruit drop;
by spraying later / to delay harvest / in citrus fruits / apples / pears;

prevent sprouting;
in potatoes; **[2x2]**

(Total 10 marks)

2

Plant growth substance	One function
	elongation growth in coleoptiles and stems / apical dominance / formation of lateral roots / delays leaf abscission / stimulates fruit development / eq.;
Gibberellins	
	ripening process in fruits / promotes abscission / eq.;
Cytokinins	
Abscisic acid	

(Total 5 marks)

Topic 3B Biodiversity

Testing your knowledge and understanding

3B.1 (a)

A	B	C	D	E
Taxonomic level	Nettle	Brooding star (Sea star)	Yeast (brewing lager)	Human
Kingdom	Plantae	Animalia	Fungi	Animalia
Phylum	Angiospermophyta	Echinodermata	Ascomycota	Chordata
Class	Dicotyledonae	Stelleroidea	Ascomycetes	Mammalia
Order	Urticales	Spinulosida	Endomycetales	Primates
Family	Urticaceae	Asterinidae	Saccharomycetaceae	Hominidae
Genus	*Urtica*	*Asterina*	*Saccharomyces*	*Homo*
Species	*U. dioica*	*A. phylactica*	*S. carlsbergensis*	*H. sapiens*

(b) Think about DNA, genes and genomes and that should point you in the direction of twenty-first century biology;

(c) (i) *brooding star and humans* – both in the kingdom Animalia (multicellular eukaryotic organisms, non-photosynthetic, with nervous coordination, heterotrophic, most able to carry out locomotion);

nettles and yeast – nettles are in the kingdom Plantae (multi-cellular eukaryotic organisms, photosynthetic, cell walls contain cellulose); yeasts are in the kingdom Fungi (eukaryotic [but yeasts are acellular rather than multicellular], non-photosynthetic, protective wall not made of cellulose [other characteristics of the Fungi do not really apply to yeasts]);

(ii) *kingdoms not represented* are the Prokaryotae (bacteria) and the Protoctista; **Prokaryotae** lack a nucleus bound with membrane, do not have membrane-bound organelles, do not have the 9 + 2 microtubule organelles, are usually unicellular (sometime filamentous), can be either autotrophic or heterotrophic; you should have plenty of possible examples; the **Protoctista** eukaryotic organisms, often but not always unicellular, not fitting into the classification of plants, animals or fungi, and a bit of a miscellaneous collection including the autotrophic (photosynthetic) algae as well as unicellular, motile heterotrophic organisms (*Amoeba, Paramecium,* etc.);

3B.2 (a)

Terrestrial habitats	Aquatic habitats
temperature	temperature
light intensity	light intensity
relative humidity	oxygen concentration
wind speed	salinity / conductivity
edaphic (soil) factors including pH, nature of particles, nutrient / humus content etc	speed of water flow / exposure (wave action)

(b) there are many different techniques, depending on chosen factors – refer to your own notes or the helpful summaries in *Tools, Techniques and Assessment in Biology,* Chapter 6, pages 91–100) and *Genetics, Evolution and Biodiversity,* Chapter 3. Reference should also be made to 'Practical work – Helpful Hints'

(c) answers depend on the habitat chosen – examples could include changes in light intensity during a 24-hour period and through different seasons in the year, or changes in salinity in a rock pool on a daily cycle linked to tidal movements and to exposure and temperature changes (evaporation of water in the pool etc.)

(d) you can compile an interesting list from your own field studies, or from textbooks – try to make sure you are familiar with some examples you have seen in the field – this means more to you and you will probably write a more convincing answer.

3B.3 use these questions as a framework for your revision. Apply them to your own records of fieldwork and the particular area(s) you studied and select techniques that are appropriate for the habitat(s) familiar to you. You will find useful guidance in *Tools, Techniques and Assessment*, by Adds, Larkcom, Miller and Sutton (published by Nelson Thornes). Look particularly at chapter 6 for details of fieldwork techniques, and at Appendix III for a summary of ways to describe vegetation. Chapter 3 outlines ways you can present data and chapter 7 will help you understand about statistical planning in your fieldwork investigation.

3B.4 (a) stage 1 cf. stage 2 – plants taller, more in flower, probably more species; stage 2 cf. stage 3 – taller shrub-like plants, probably different species; stage 4 cf. stage 3 – tall trees, different ground flora, fewer shrubby plants;

(b) mowing; grazing (e.g. rabbits, sheep etc.); burning; eq.; plagioclimax (if seral succession prevented or deflected);

(c) temperate climate, say 20 to 30 years (but relate your answer to a situation that is familiar to you);

(d) terms could include succession / seral stages – final stage is climax community (at stage 4);

(e) depends on the location and other factors, but usually greatest diversity at the middle stages – so probably stage 3 (or stage 2); try to give some plant names from your own school or college grounds or another area you may have visited; animal species also likely to change, different niches / habitats / opportunities / food chains etc.;

(f) ground flora growing and flowering in spring, when more light available (before leaf canopy develops); other answers could refer to coppiced woodland (you could check back on your AS notes about coppiced woodland) and Qu 3B.10 in the conservation section;

3B.5 *open water to edge of pond* – succession from aquatic plants and animals etc adapted to living in open water to terrestrial organisms, including those at edge where water depth is becoming progressively less; climax likely to be woodland (*Note*: but not in all localities – can you suggest why and where not?); prevent by clearance, particularly at edge to maintain open water;

sand dune – succession from almost bare loose sand with plants able to colonise and tolerate the shifting substratum with associated animals etc which are adapted to living in unstable situation to more permanent communities with organisms which show preference to the more stable situation; climax likely to be woodland (*Note:* but not in all localities – can you suggest why and where not?); prevent by maintaining blown-out areas of sand or a means to prevent colonisation and establishment of a stable community;

3B.6 (a) (i) population; community;
(ii) numbers; density;

(b) show lag, exponential, stationary and death phases (*Note:* for growth curve, the lag phase shows only a small increase in the number of organisms, whereas in the exponential phase there is a steep increase in numbers and sharp rise in the curve. On the curve, the stationary phase is horizontal followed by a fall in the death phase, due to decline in numbers. This curve is shown in *NAS Genetics, Evolution and Biodiversity*, page 47, Figure 4.2); factors include nutrient supply, availability of oxygen, presence of toxic substances, etc.; (*Note*: link this with effects such as eutrophication)

(c) might get similar but fluctuating curve; time scale depends on generation time of the organism, as well as other factors such as availability of suitable food supply, presence of predators, etc. (this curve is shown in *NAS Genetics, Evolution and Biodiversity*, page 48, Figure 4.4);

(d) 1 = density; 2 = time; 3 = potential population curve; 4 = environmental resistance; 5 = population curve; 6 = carrying capacity;

(e) *carrying capacity:* the optimum size of population that can be supported in a given habitat or area;

environmental resistance: the environmental factors (such as competition, food availability, disease and space) which restrict or limit the growth of the population;

(f) *nutrients* – increase generally allows increase in population numbers; link with eutrophication;

light – link with photosynthesis (diurnal changes as well as seasonal changes);

temperature – link with effect on enzyme reactions – including photosynthesis – (diurnal fluctuations as well as seasonal changes);

(g) *intraspecific* – refers to competition between members of the same species, for factors such as nutrients, water, light etc – relevant to spacing of crop plants – closer spacing may lead to lower yields;

interspecific – refers to competition between members of different species, so is relevant to competition between weeds and crop plants (again for available nutrients, water supply, light etc); think of other examples involving animals – in relation to feeding, shelter (protection) etc.;

(h) the terms *predator* and *prey* generally refer to a feeding relationship between animal species, in which the predator is likely to kill the prey before eating it; the curve fluctuates showing a series of peaks and troughs, out of phase, in which the peak of the prey is followed by a peak for the predator and usually the peak for the prey is lower than that for the predator (this curve is shown in *NAS Genetics, Evolution and Biodiversity*, page 50, Figure 4.9); link to biological control of insect pests in which the pest is the prey and the biological control agent is the predator – the fluctuations in the curves should also suggest to you how the control organisms are applied to the pest;

(i) *immigration* (from other localities) – increase in population numbers;

emigration (to other localities) – decrease in population numbers;`

disease – may also increase the 'death rate';

3B.7 (a)

Feature of control method	Chemical	Biological
Action general – can kill wide range of insects	✓	✗
Action often specific – targets particular species or group of species	✗	✓
Often does not completely eliminate the pest	✗	✓
Toxicity may affect other (non-pest) animal species	✓	✗
Acts relatively slowly	✗	✓
May persist over long period – spread through food web etc	✓	✗
Effectiveness may diminish as numbers of resistant strains increase in the population	✓	✗

(b) may need to release at certain stage of life cycle of pest; may be linked to suitable environmental conditions – for pest or control agent (e.g. temperature); if pest eliminated, predator would have no food and so not persist through to another attach by population of pests;

(c) chemical pesticides likely to be more costly – expensive because of expensive research and development programmes, manufacturing process, machinery to apply, need to repeat applications, likely to eliminate natural predators;

biological control methods less expensive – less in terms of development programmes, may be some costs in rearing and distributing the predators, simpler to apply and may persist over long period (if the population survives) thus avoiding need for repeated applications;

(d) *beetle banks* – strips in and near crops, often planted / sown with appropriate vegetation, allow overwintering of natural predators which then likely to attack pests on growing crop, spacing and distribution adjusted to be right for distance predators likely to travel when foraging in the crop;

Phacelia tanacetifolia – sometimes grown along margins of cereal crops, their blue flowers attract hoverflies, hoverfly larvae are predators on aphids;

intercropping – increases biodiversity, increases likelihood of natural predators; (see Question 3B.12 on conservation practices within farmland)

(e) practices that encourage natural predators include: maintaining hedges, ponds etc within an area ; timing of certain agricultural practices, such as mowing, or ploughing in stubble ; rotation of crops ; undersowing of one crop with another (e.g. cereal crop undersown with a legume) [this has a similar effect to intercroppping];

(f) combination of approaches to pest control, including chemical, biological and cultural methods (see *Genetics, Evolution and Biodiversity*, chapter 4);

(g) reference to persistence over long period and spreading through food web; accumulation of DDT is an often quoted example – try and find some others;

3B.8 maintain range of habitats; protect existing ecosystems; maintain and increase species diversity; maintain genetic resources; protect rare / endangered species (or perhaps provide the necessary habitat so that they survive); leisure benefits; conserved areas / nature reserves provide opportunities for study and research; reasons for – change of land use leading to loss of habitats (buildings, roads, agriculture etc.);

3B.9 (a) mowing, grazing, scrub clearance (later in succession); (*Note:* see also other answers, for example Questions 3B.4 and 3B.5, in relation to maintaining a range of habitats, hence diversity.)

(b) open water gets filled up / silted / accumulation plant debris / eq.; plants appropriate for shallow water (and associated fauna) can then colonise progressively to dry land; keep at least some clear open water (dig out silt etc and remove debris) + gradual slope at edge to encourage range of plants etc; likely to increase species diversity (because maintaining range of habitats);

3B.10 maintains range of habitats (successive stages of coppicing + rides or paths between coppiced areas) + encourages species diversity; other benefits also evident – protection rare species, leisure activities, area for study, not converted to farmland, etc;

3B.11 moorland (UK) – e.g. heather dominated – helps keep plants of different heights, good for range of grazing animals + fauna (including birds) associated with / preference for different stages of regeneration of plant cover; also useful in tropical / semi-tropical forest – maintains / increases habitat diversity;

3B.12 (a) small-scale intercropping; intercropping has another crop between the main crop – perhaps different height or to be harvested at a different time – allows greater diversity of associated fauna (insects etc.) and the ground cover provided by one crop may help suppress weeds; monoculture has vast swards of same species so little scope for diversity of fauna etc – often chemicals used to control weeds and pests;

(b) very few are 'natural' if we define 'natural' as having had no human influence – grasslands often grazed by domesticated animals, in woodland timber products are harvested etc – probably inaccessible extreme areas are the nearest we can get – but we tend to use 'natural' in the sense of not farmed (intensively) so woodland, hedges, marshes, ponds etc can (and are) treated as 'natural';

(c) many examples could be quoted here, depending on the type of farmland – some include – minimal use of fertiliser, pesticide and herbicide; management of grassy areas (as above) by grazing, sensible stocking of animals (so not over-grazed, but allowing trampling by cattle, different animals with different species preference); maintenance of hedges in stages and keep some trees to allow diversity; management of water courses and ponds (as above); use of 'beetle banks', corridors of uncultivated land, headlands etc.; allow diversity in verges to lanes within farmland, timing of cutting to allow flowering; etc.;

3B.13 the Habitats Directive aims to create a network of special ecological areas of conservation within Europe. Other points include: recognition that natural habitats are continuing to deteriorate; there is a consequent threat to (loss of) species of animals and plants; provision of a list of habitat types that need to be designated as special areas of conservation;

then use your own knowledge to draw up a list of points relating to legislation – what it aims to do and any specific examples relating to actual habitats; Make sure your list refers to actual (or possible) examples of successes or failures and is not just a vague set of ideas based on your opinions. In your examples of legislation, you should be aware of those that apply at a local, national as well as global level. Remember that it is an understanding of the biology of the area that is important in drawing up any strategy for conservation;

Practice questions

Unit 5B: Practice questions

1 (a) *L. littorea* has greater range / *L. saxatilis* has narrower range / eq.;

L. littorea found lower on shore / *L. saxatilis* found higher / eq.;

L. littorea more abundant / *L. saxatilis* less abundant / eq.;

both occur at site 8 and 9 / ref. to overlap;

neither occur at sites 1 and 2 / on lower shore;

L. saxatilis most abundant at sites 10 and 11, *L. littorea* at site 7; **[3]**

(b) Rough periwinkle / *L. saxatilis*;

will be out of water / exposed for longer / eq. / less time in water / allow converse for *L. littorea*; **[2]**

(c) food availability / food distribution; predation; parasites / disease; conditions for reproduction;

temperature; competition; salinity; difference in wave action / tidal power; **[2]**

(Total 7 marks)

2 (a) fall / decrease; increase in concentration needed each year (to kill some number of pest); credit for correct use of figures (need some manipulation / % decrease calculated / eq.); **[2]**

(b) a few / some individuals in population are resistant / eq.;

resistance due to an allele / gene / eq.

ref. to mutation as origin of resistance;

resistant individuals survive to breed;

resistance passed on to offspring;

so more survivors / fewer killed in next generation;

possibility of more than one gene for resistance so combination of genes may increase resistance;

reference to natural / artificial selection / selection pressure / survival of the fittest; **[4]**

(c) (i) insecticide / eq. is persistent / non-biodegradable / eq.

small / eq. amount taken up / absorbed by pest / eq.;

stored / retained / not excreted by organisms;

many 1° consumers eaten by 2° consumers, so 2° consumer receives more,

animals at end of food chain / eq. receive most / eq. named e.g. / ref. to DDT and fat solubilty; **[3]**

(ii) can be made specific / selective / allow description,

non-persistent / biodegradable/eq.; **[2]**

(Total 11 marks)

3 (a) no. of avocets rises/eq. or figures quoted to 1957; then avocet numbers fall and then rise to 1961; gull numbers fluctuate /eq. until 1955; then gull numbers rise and fall until 1961; **[3]**

(b) (i) 3600–700 (from graph) / $3.6 \times 10^3 - 0.7 \times 10^3$;

$\dfrac{2900 \times 100}{3600}$;   80.6 % / 80.56 % / 80.5 %; **[3]**

(ii) population increases (rapidly) / almost doubles / figs. quoted;

due to reduced / less predation eq. by gulls; **[2]**

(iii) changes in food availability;

changes in named climatic factor

high / low rainfall levels;

(still some) predation / eq. by gulls;

intraspecific competition / OR competition in named factor e.g. nest sites / land / space; effects of other predators; qualified ref. to pollution;   [2]

(c) mowing, periodic cutting of grass; decrease in coarse grass/stops succession, eq.;

flooding / drainage / eq.; controls water levels, maintains species, eq.

coppicing; cut trees when young / eq.; increase ground flora/eq.;

other suitable examples e.g. grazing / use of pesticide / culling;   [3]

(Total 13 marks)

Topic 4B Genetics and evolution

Testing your knowledge and understanding

4B.1 false – alleles are different forms of a gene;

4B.2 (a) Bb;

(b) bb;

4B.3 phenotype;

4B.4 ABO blood groups;

4B.5 *autosomal linkage:* the association of two or more genes; on the same (non-sex) chromosome;

4B.6 males;

4B.7 (a) height / weight / intelligence;

(b) sex / eye colour / ABO blood group;

4B.8 (a) discontinuous;

(b) continuous;

4B.9 meiosis;

4B.10 they introduce *new* genes, rather than simply mixing up the current gene pool;

4B.11 (a) nitrous acid / 5-bromouracil;

(b) UV light / X-rays;

4B.12 deletions, insertions and substitutions;

4B.13 *translocation:* the movement of one part of a chromosome to another, non-homologous chromosome;

4B.14 (a) stabilising selection;

(b) directional selection;

(c) disruptive selection;

4B.15 speciation;

4B.16 geographical isolation and behavioural isolation;

4B.17 (a) endonuclease;

(b) DNA ligase;

(c) reverse transcriptase;

4B.18 a small, circular loop of DNA separate from the circular chromosomal DNA of bacteria;

4B.19 small size makes them easy to handle; able to replicate within bacterial cells;

4B.20 insulin / human growth hormone / erythropoietin / blood clotting factors;

Practice questions

1

Phenotype	Genotype
Agouti	AA, AC, AH, AW;
Chinchilla	CC, CH, CW;
Himalayan	HH, HW;
Albino	WW;

(Total 4 marks)

2 (a) Aa   aa;   [1]

(b) (i) genotype – AaEe;

phenotype – purple;   [2]

(ii)

	AE	Ae	aE	ae
AE	AAEE	AAEe	AaEE	AaEe
Ae	AAEe	AAee	AaEe	Aaee
aE	AaEE	AaEe	aaEE	aaEe
ae	AaEe	Aaee	aaEe	aaee

[3]

(iii) 9 purple;

4 white;

3 red;   [3]

(iv) aaee;

Aaee;

AAee;   [3]

3 (a) monohybrid involves one character, dihybrid two characters/genes/eq.;

monohybrid one gene locus, dihybrid two loci/monohybrid one pair of alleles, dihybrid two pairs of alleles involved;

credit for example of each;   [3]

(b) continuous has complete range of phenotypes, discontinuous only a few categories/eq.;

continuous polygenic/controlled by large number of genes, discontinuous one/only a few genes;

continuous likely to/may have large environmental influence, discontinuous little/no environmental influence;

credit for examples of each; **[3]**

(Total 6 marks)

4 (a) refer to genetic isolation between members of one population / eq.; or prevents breeding between populations / species.

need both → postzygotic – after mating eq. / fertilisation, prezygotic – before mating eq. / fertilisation; (NOT just zygote)

postzygotic example-(hybrid) inviability, won't develop to maturity / (hybrid)

sterility or infertility ref. / eq.;

prezygotic example- temporal isolation (time / season) / eq. / different

reproductive structure (incompatible genitalia).

/ different reproductive behaviour (mating / courtship) / eq.; (NOT geographical isolation) **[3]**

(b) speciation means divergence or formation / eq. of two or new (or more) species / eq.;

need both → sympatric – without physical, geographical / eq separation,

allopatric- after physical, geographical / eq. separation;

sympatric example described eg development of polyploidy in plants / herring

gulls / cabbages and radishes / eq.;

allopatric example described e.g. Darwin's finches; **[3]**

(Total 6 marks)

5 (a) (i) (restriction) endonuclease / restriction enzyme / named example;

e.g. HIND 3

EcoRI **[1]**

(ii) by its base sequence;

related to amino acid sequence;

use of a (gene) probe / gene marker;

credit further details of probe, e.g. complementary or radioactive;

insert (gene) into bacteria;

select bacteria which produce T-toxin; **[2]**

(b) reference to use of a plasmid / virus / gene gun / eq.;

as a vector / refer to use of protoplasts with gene gun;

reference to use of ligase to attach gene to plasmid;

treatment with calcium ions / heat treatment to make cell membrane permeable

(heat shock / electric shock);

reference to *Agrobacterium*; **[2]**

(c) transcription of gene (for T toxin) to form mRNA / description of process;

mRNA attaches to ribosomes;

mRNA codons translated to sequence of amino acids;

tRNA brings appropriate (specific / particular) amino acids / eq.;

amino acids joined by peptide bonds; **[4]**

(d) advantage – no need to use insecticides / environmental benefit – qualified / increased yield – qualified / refer to killing insects which eat the crop;

disadvantage – consumer preferences / insects may become resistant to T-toxin / possible to adverse effect on other organisms / may affect the taste; / gene may cross into another organism. **[2]**

(Total 11 marks)

6 (a) gene is a specific sequence/length of DNA/occurs at a locus on a chromosome;

with a specifc function/codes for a particular polypeptide/protein responsible for a characteristic;

alleles are alternative forms of a gene;

only one allele present at a locus/alleles are separated in meiosis; **[3]**

(b) (i) $1=I^OI^O$; $2=I^AI^B$; $4=I^AI^O$; $5=I^BI^O$; $6=I^AI^O$; **[5]**

(b) (ii) could be either A or B/AO or BO/eq.; would inherit either A or B from mother/female/one parent; but O only from father/male/other parent/ other parent is homozygous for O; **[3]**

(Total 11 marks)

Unit 5H Genetics, human evolution and biodiversity

Topic 1H Genetics and evolution

Testing your knowledge and understanding

1H.1 amniocentesis; chorionic villus sampling;

1H.2 *karyotype:* the number, type and structure of the chromosomes of an individual;

Practice questions

1 (a) 47; **[1]**

(b) (chromosome in bottom right-hand corner); **[1]**

(c) a condition in which an organism has one more chromosome than normal; **[1]**

(d) female gamete contains two copies of chromosome 21; male gamete contains one copy of chromosome 21; the gametes fuse at fertilisation to produce an individual with three copies of chromosome 21 (trisomy); which causes Down's syndrome; **[3]**

(Total 6 marks)

Topic 2H Human evolution

Testing your knowledge and understanding

2H.1 chimpanzee has fewer teeth; ridged molars; larger canine teeth / incisors; a brow ridge; deeper upper jaw;

2H.2 lemurs;

2H.3 immunological studies / antigen-antibody reaction; amino acid sequencing in proteins; DNA profiling;

2H.4 palaeontology; comparative anatomy;

2H.5 *geochronology*: the methods used to find the age and sequence of geological events;

2H.6 Hominoids – Australopithecines – *Homo habilis* – *Homo erectus* – *Homo sapiens*;

2H.7 modern humans have: no brow ridge; a flattened face; a larger cranium;

2H.8 *H. habilis* did not use fire; used Oldowan / pebble tools; did not produce cave art; did not bury dead;

2H.9 (a) Neolithic;

(b) 10 000 years BP;

2H.10 making tools; sharp edges for knives / saws; killing animals; preparing skins / eq.;

Practice questions

1 (a) B, C, A **[1]**

(b) any THREE from: brow ridges (less prominent); cranium (more domed); face (flatter); teeth (smaller / flatter); (upper) jaw (shallower) / zygomatic arch; **[3]**

(Total 4 marks)

2 (a) have a particular shape / eq. / fit a particular molecule / lock and key idea; **[1]**

(b) significantly / eq. different structure / greater difference / eq. / distant relationship; refer to lower % precipitation / less good fit / eq.; **[2]**

(c) (i) Tree B; chimpanzee protein more similar to human than gorilla / suitable use of figures, e.g. 3 % difference; **[2]**

(ii) branch should be drawn on Tree B from gibbon line; **[1]**

(d) similar proteins show similar DNA (sequence) / genes / eq.; (DNA) code is used to synthesise proteins / template ref.; similar DNA implies close relationship; **[2]**

(e) primary protein structure / amino acid sequence directly determined by DNA / eq.;

structure rather than sequence important for antibody method / depends on efficiency of immune system / eq.;

small change in sequence could give large change in structure / converse;

amino acid sequence should show DNA changes more accurately;

reference to use of another animal is less accurate / eq.; **[3]**

(Total 11 marks)

3 *habilis*; *erectus*; fire / weapons / spears; 250 000 BP / 0.25 million years ago (allow anything between 130 000 to 400 000); baskets / basketry / jewellery / carving; Neanderthal man / *Homo sapiens neandertalensis*;

(Total 6 marks)

Topic 3H Human populations

Testing your knowledge and understanding

3H.1 (a) dip almost certainly related to events in China (where population fell by about 13.5 million – massive loss of life during the great famine of 1960–1962, linked to series of droughts and floods); other suggestions could include large-scale loss of life through wars, famine for a variety of reasons, or other events;

Examiner's comments

Events must be large scale as you are looking here at world population.

(b) times of war (e.g. 1914–18, 1938–45); loss of population due to people being killed + mainly men fighting so lower birth rate for women left at home;

(c) mid- to late twentieth century mainly 'developing' countries; last two centuries, mainly times of European expansion, into colonial countries and Americas;

(d) China – 1 273 111 290; UK – 59 647 790;

(e)

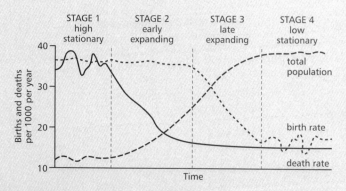

The demographic transition model of population growth. The terms 'high stationary phase' and 'low stationary phase' refer to the birth and death rates rather than to the numbers in the population.

3H.2 (a) horizontal axis = percentage of population (each side of central axis which is designated '0'); horizontal bars represent population (like a bar chart drawn on its side); one half of pyramid = males, the other half = females (this curve is shown in *NAS Genetics, Evolution and Biodiversity*, page 151, Figure 11.8);

(b) narrower at the top because relatively smaller and smaller proportion of the population live to old or very old age; pyramid with very wide base represents developing country where birth rate likely to be high and deaths in childbirth and early years of life;

other events, which would alter relative length of horizontal bars, include war and famine, migrations to other countries (often pushed by economic pressures, but not relevant in a *world* population), disease, social or governmental policies (e.g. imposition of one-child policy in China which would reduce bars (in relation to earlier age cohorts);

3H.3 (a) *fertility* = number of live births in a population over a period of time; fecundity = the physiological ability of a woman to conceive and bear children (can also be applied to a population ;

crude birth rate = total number of live births in one year × 1000 divided by the total mid-year population; *crude death rate* = total number of deaths in one year × 1000 divided by the total mid-year population;

(b) *general fertility rate:* includes the whole population, including those outside reproductive age and those unable to conceive (single people or with reproductive abnormalities which makes them infertile);

total fertility rate: relates more closely to the number of women of child-bearing age (taken as 15–49 years);

Examiner's comments

The list below, for (c), gives a group of 'social' factors, which have changed over time, and are different in different countries – it is worth being familiar with some examples.

(c) *proportion of women in marriage* – higher proportion likely to give higher fertility rates; note also influence of cultural factors relating to birth outside marriage;

age of marriage – general trend for later marriage (cf. primitive societies), which is likely to delay age at which first child is born and reduce overall fertility rate;

family size – traditional families were often large – modern society favours smaller families, linked with use of birth control (personal or religious decisions and government imposition – e.g. China's one-child policy), women preferring to establish careers before rearing family, changes in agricultural practices which reduce dependence on family and child labour etc.;

(d) traditional families (lower fecundity compared with modern families) – later age of first menstruation; delay of ovulation even when menstruation has started; suppression of menstruation and ovulation during breast-feeding (lactational amenorrhoea);

modern family uses methods of contraception – e.g. barrier methods, hormonal methods, interference with implantation; China's one-child policy (already referred to) with penalties for those who do not comply; Romania (during 1960s) had strong legislation to encourage births, including anti-abortion legislation, banning the sale of modern contraceptives; government legislation to encourage or discourage women working (through financial incentives or tax system); economic pressures on women to work;

3H.4 (a) possible reasons are linked to pollutants, particularly oestrogens and other synthetic chemicals that mimic hormones and may interfere with the development in the uterus or early in life; other factors include keeping the testes at too high a temperature, stress, tiredness, smoking, caffeine, alcohol;

consequences – likely to lead to lower fertility rate (even though only one sperm is required to fertilise a single ovum);

(b) in *women* – failure to ovulate, obstruction of the oviduct(s), diseases of the endometrium, cervical mucus not receptive;

in *men* – lack of sperm, decrease in number of normal sperm, decreased motility of sperm, blockage or abnormality in male tract;

in *both* – poor hygiene and sexually transmitted disease tend to give reduced fertility;

(c)   *in vitro* fertilisation (IVF), artificial insemination (AI); (can you work out the meaning of different acronyms associated with these methods of 'assisted reproduction' – GIFT, ZIFT, DI, HI);

Practice questions

1 (a) (i)   increasing: any TWO countries from: Costa Rica / Jordan / Thailand;

stable: any TWO countries from: Bulgaria / Nepal /Korea / Romania;   **[2]**

(ii) contraception / infertility / fertility treatment; religion; marriage / age of marriage / family size / culture / eq.; economic factors / wealth / war / food supply / famine / medicine / diet); education; government policies / eq.;   **[3]**

(b) death rate / mortality / immigration / emigration / public health / diet / war / disease / disaster / government policies / economy / wealth;   **[1]**

(Total 6 marks)

2 (a) number of (live) births / individuals being born per number of the population / 1000 / eq. / in a given time;   **[2]**

(b) more people living beyond 65 / greater life expectancy, in 2020; better health care;

(much) greater numbers of people between 15–64 years in 2020;

better health care / fewer die in childhood / better diet / food supply / large number of children 0–14 in 1990 / eq.;

numbers of people in each age group between 0–30 / 0–14 almost the same in 2020;

better health care / fewer die in childhood / better diet / eq.;

more women in all age groups in 2020;

fewer deaths due to child birth / better screening for women;

more people in all age groups in 2020; better health care / fewer die in childhood / better health care / eq.

more babies predicted to be born in 2020;

better health / diet etc.;

[2x2]

(Total 6 marks)

3 (a) 1891; refer to widest base / highest % children / eq. in the population (0–9 age range);   **[2]**

(b) 1891 females equal to males but in 1947 females greater than males;

1891 smaller % in each age range / stated age range / refer to figures;

1891 gradual / eq. decrease at each age range but 1947 increase in 35–9;

1891 poorer hygiene / public health / converse for 1947;

1981 less medical provision / eq. / converse for 1947;

1891 poorer diet / food availability / food production / converse for 1947;

1947 fewer men due to war / converse for 1891;   **[4]**

(Total 6 marks)

Unit 6 The GCE Advanced Assessment

Synoptic paper

Practice questions

1 (a) sodium re-enters the blood / is reabsorbed; at same rate as water; urea is not / less urea returned to the blood / more urea stays in the tubule;   **[2]**

(b) sodium ions – must be reabsorbed into the blood; mostly in distal convoluted tubule; potassium ions – secreted into tubule; water absorption from tubule increases concentrations of ions / eq.;   **[3]**

(c) active transport; transport proteins / reference to Na / K pump / ATP-ase activity / co-transport with Na; correct reference to diffusion initially;   **[2]**

(d) X = mitochondrion; provides energy / ATP / reference to aerobic respiration; Y = microvilli / brush border; increases / provides larger surface area for reabsorption;   **[4]**

(e) nucleus / nuclear membrane; endoplasmic reticulum; lysosome; Golgi body;   **[2]**

(Total 13 marks)

Examiner's comments

- Think very carefully about the answers to (a) and (b), to make sure that you answer relevantly. Look at the data and think about the water, which is not included in the table of data.
- The answer to (c) requires some knowledge from Unit 1.
- The answer to (d) also requires you to recognise features in cells (Unit 1) and to relate this to some knowledge gained in Unit 4.
- The answer to (e) relates to Unit 1.

2 (a) Cell B = mitosis + same number of chromosomes as A; Cell C = meiosis + chromosome number halved / reference to crossing over;   **[2]**

(b) pairing of homologous chromosomes occurs; chiasma formed / crossing over took place; exchange of genetic information between chromatids; **[3]**

(c) dominant; rats are resistant to warfarin in the heterozygous condition; **[2]**

(d) W_1W_1 is killed by warfarin / not resistant; W_2W_2 needs so much vitamin K; **[2]**

(e) homozygous recessive no longer at a disadvantage; so numbers of non-resistant rats will increase; homozygous dominant still needs high quantities of vitamin K; greater competition for food resources in environment; no effect on heterozygotes; correct reference to change in selection pressures; **[3]**

(f) resistant alleles still present in heterozygotes; may occur as a mutation; no selection against heterozygotes / reference to the need for many generations before allele frequency changes; **[2]**

(Total 14 marks)

Examiner's comments

- Be very careful about the answers to the first two parts of this question as it is all too easy to get the answers the wrong way round.
- If you read the information carefully you should have found the answers to some parts easily. If you did not get them right, or could not think of an answer, go back and read all the information again.
- Be prepared for unfamiliar data and situations. You might not have heard of rats and warfarin resistance but you should be able to apply your biological knowledge to this situation and come up with reasonable explanations. Sometimes there are no 'right' answers, or you would not be expected to know the exact answer, but you will be awarded credit for sensible suggestions.

Written alternative (W2) to the individual study

Practice questions

1 (a) suitable table; correctly labelled rows and columns; loss in mass correct for species A; loss in mass correct for species B; **[4]**

(b) suitable axes labelled; correct plot for A; correct plot for B; line graphs, correct key; **[4]**

(c) species B loses water more rapidly than species A; both rates slow with time / overall pattern similar; **[2]**

(Total 10 marks)

Examiner's comments

All this is very straightforward as long as you remember to check your calculations and make sure you plot the graph with the axes the right way round.

2 (a) There are plenty of possible marks here, depending on the method you decide upon, but you would need to include reference to the following:

same variety of plant used / genetic uniformity; same age of plants in experiments; organisation of test plots; stated number of plots; half infected with aphids; methods of clearing aphids; standard conditions maintained in glasshouse; standard treatment of plants throughout test; length of time of experiment; how results judged / stated parameter measured; **[9]**

(b) tabular form; suitable columns and rows labelled; means calculated; bar chart / eq. according to method; correct axes; null hypothesis stated; use of t test; reference to confidence limits; **[7]**

(c) Again, there are many possibilities here, but you would be expected to refer to the difficulties of infection, of keeping plots aphid-free, the effects of insecticides on the growth of the peppers, reference to other diseases. Further work could include effects of different numbers of aphids, effectiveness of biological control or effects at different stages of the life cycle of the plant. **[5]**

(Total 21 marks)

Answers to assessment questions

Unit 4 Respiration and coordination and options

Core section (Topics 1 and 2)

1 **(a)** pyruvate / pyruvic acid; **[1]**

(b) 2 (molecules); **[1]**

(c) $2 \div 36 \times 100 / 0.056 \times 100$;

5.6 / 5.55; **[2]**

(Total 4 marks)

2 **(a)** A photoreception / eq. / correct reference to rhodopsin / visual purple;

B synapse with *bipolar* neurone / release transmitter substance; **[2]**

(b) (cones / receptor cells) most tightly packed (in fovea);

(mostly) cones in fovea / (mostly) rods away from fovea;

cones have greater acuity / resolution than rods;

each cone sends information separately / no convergence / has its own bipolar cell / neurone;

rods pool information / convergence / share bipolar cells / share neurones;

light passes through fewer blood vessels / bipolar cells ganglion cells (at fovea); **[3]**

(Total 5 marks)

3 **(a)** A Bowman's capsule / renal capsule;

B 1st / proximal convoluted tubule;

C loop of Henle;

D 2nd / distal convoluted tubule; **[2]**

(b) **(i)** water (re)absorbed; by osmosis / following (re)absorption of a solute; urea not (re)absorbed / limited (re)absorption; **[2]**

(ii) *varying* amount of water reabsorbed;

refer to ADH;

increases permeability of DCT / collecting duct to water;

(ADH) from *posterior* pituitary;

refer to concentration / water content of blood / osmoreceptors / baroreceptors affects ADH secretion;

refer to varying protein diet; **[3]**

(c) **(i)** $(0.36 - 0.30) \div 30$;

= 20 %;

increase; **[3]**

(ii) more excess amino acids;

deaminated;

in the liver;

so more ammonia / ammonium ions produced;

converted to urea;

therefore more urea transported to kidney in blood / more urea filtered from blood; **[3]**

(Total 13 marks)

4 **(a)** potential (difference across a cell membrane) is reversed / eq. reference to depolarisation; +40 mV; inside becomes more positive / converse; by opening of Na$^+$ gates / increase in permeability of membrane to Na$^+$ / Na$^+$ moves in; passes along / moves along / propagated along /reference to wave; **[3]**

(b) Schwann cells cover / wrapped around axon /eq.; contain lipid phospholipid fatty material reference to Schwann cell membrane; electrically insulates axon / eq. non-conductive; reference to nodes of Ranvier / axon exposed at certain points enables saltatory conduction / 'jumping' of action potentials impulse increases faster conduction of impulse **[3]**

(Total 6 marks)

5 The following points would gain credit:

rod-shaped organelles in cytoplasm of eukaryotic cells; up to 1 mm wide and 7 mm long; surrounded by an envelope / double membrane; outer membrane smooth; inner membrane folded into shelf-like structures / cristae; inner membrane encloses the mitochondrial matrix; they are the sites of aerobic respiration; matrix contains enzymes involved with the tricarboxylic acid (TCA) cycle / Krebs cycle; pyruvate produced from glycolysis passes into matrix; description of link reaction / pyruvate converted into acetate; combines with coenzyme A to form acetyl coenzyme A; NADH and carbon dioxide formed; four-carbon oxaloacetate combines with acetyl unit from acetyl coenzyme A to form a six-carbon compound / citrate; sequence of reactions occurs in which citrate converted back to oxaloacetate; catalysed by enzymes in the matrix; carbon dioxide is released and NAD and FAD are reduced; these reduced electron carriers are reoxidised in the etc / electron transport chain; free energy is liberated which is used to generate ATP; this process / oxidative phosphorylation takes place on the cristae; details of involvement of cytochromes;

(Total 10 marks)

Option A Microbiology and biotechnology

1 (a) (i) $(1700 - 900) \times 100 \div 1700$;

47 %; **[2]**

(ii) better drugs / medicines / medication AZT / new drugs available;

to delay onset of AIDS / eq. (in HIV positive people) / eq.; **[2]**

(b) refer to virus, *latent period* / *latency* / eq.;

infected (T4) cells need to be activated / triggered / eq. before replication of the virus takes place;

to cause further destruction of (T4) cells;

AIDS only develop[s after depression of immune system / activation by secondary ? eq. infection / or reference to natural immunity to AIDS;

(therefore people) can carry HIV without any illness / symptoms *or* may die before AIDS develops / eq. **[3]**

(c) no symptoms / eq. with infection;

(therefore) not detected unless a blood / tissue / saliva / DNA test taken; **[2]**

(d) carries (enzyme) reverse transcriptase;

synthesises DNA using (viral) RNA;

DNA codes for synthesis of viral proteins / inserted into host genome / inserted into host's chromosomes / host DNA; **[2]**

(Total 11 marks)

2 (a) Reference to presence of (raw sample contains) a very large number of bacteria; *colonies* would be too many to count *accurately* / eq.; *colonies* would be too close together / merge / overlapping; **[2]**

(b) Reference to use of sterile pipette / sterile / aseptic technique; add to (sterile nutrient) agar (in petri dish); mix / eq. and allow to set / description of spreading (*not* streaking); use different / new sterile / eq. pipette / spreader for each sample; mix / homogenise contents of the tube before removing sample; **[3]**

(c) (i) Use plate *D*; has a reasonable / eq. number of colonies; Plate C has too many / too densely populated / too close together / overlapping colonies; Plate E has too few colonies / not reliable / eq.; **[3]**

(ii) Number of colonies on Plate D = 23; Number of colonies in 1 cm^3 of D = 230 / number of bacterial cells in original sample = $230 \times 1/10^{-4}$; = 2 300 000/2.3 ×106; **[3]**

(d) Dead cells not counted / only counts viable cells; **[1]**

(Total 12 marks)

3 (a) to kill (all) bacteria / microorganisms; refererence heat-resistant bacterial spores / thermophilic species; **[2]**

(b) to prevent the cotton wool being soaked/ getting wet; **[1]**

(c) flaming of inoculating loops / forceps / eq. bottle neck; kills microorganisms present on surface; introduce loops into flame slowly; to avoid microbial aerosol formation / spluttering / eq.; swabbing laboratory work surfaces with disinfectant / eq.; reduces numbe of *contaminating* organisms; reference to covering / sealing / minimum opening of dishes / cultures; reducing entry of / exposure to air; avoid breathing over cultures / eq.; reference to human contamination; **[4]**

(Total 7 marks)

4 (a) *Fusarium (graminearum)*; **[1]**

(b) ammonia as a *nitrogen* source; required as a component of *protein* / *amino acid* / nucleic acids (molecules); air, as source of oxygen / for aerobic conditions; for respiration / fermentation; **[3]**

(c) rapid / sudden / eq. rise in temperature *or* refer to thermal shock; denature proteases / inactivate proteases or breaks down / eq. RNA; **[2]**

(d) respiration / fermentation / metabolic activity / chemical reactions of the fungus generates / releases heat;

(heat exchanger) cools culture / reduces temperature / removes heat;

refer to maintenance of optimum temperature for growth; **[2]**

(Total 8 marks)

Option B Food Science

1 (a) resazurin tests for the freshness of milk / effectiveness of pasteurisation, turbidity tests for the effectiveness of sterilisation; resazurin shows (metabolic) activity of bacteria / detects live bacteria, turbidity detects changes / presence of proteins; rezaurin is blue (in oxidised state), changes to white / pink / mottled if (quantities of) bacteria are present; turbidity: if sterilisation is effective / if no live bacteria present, filtrate should remain clear; / eq.; add ammonium sulphate for turbidity test; resazurin in resazurin test **[3]**

(b) freezing involves storage at very low temperatures (at least –18 °C / –20 °C), canning is a form of heat sterilisation / eq. very high temperatures used / heated to around 115 °C; freezing prevents microbial growth / retards enzyme action / activity of microorganisms, canning kills all microorganisms; keeping times longer for canned foods than frozen foods / eq.; more / greater loss of nutrients / vitamins / changes in texture in canned foods / /texture usually the same / less loss of vitamins in frozen foods; canning creates / produces anaerobic conditions / always excludes air **[3]**

(Total 6 marks)

2 (a) muscle wastage / loss of muscle / thin limbs / protruding bones; lack of fat / adipose tissue; *grossly* / eq. underweight / emaciated / *very* thin; old (man's) wizened face / wizened face; **[2]**

(b) diet low in calories / energy *and* protein / protein-calorie malnutrition; due to early weaning; on to insufficient / too dilute food / starvation idea; poor hygiene leads to gastro-enteritis / diarrhoea; body protein / muscle used as an energy source / eq.; **[3]**

(Total 5 marks)

3 (a) Any TWO of: rice / white bread / new potatoes / pasta spirals / rolled oats; **[2]**

(b) need carbohydrate for energy; but need to avoid sugars in food / these foods low in / no sugar / low GI so release sugars slowly; also need some fats; preferably as unsaturated fats / these foods contain unsaturated fats / no/low in saturated fats; reference to suitable foods have high starch content; **[3]**

(Total 5 marks)

4 (a) wheat / flour / maize meal / rice / potatoes / eq. any reasonable starchy material; **[1]**

(b) bacteria and / or moulds / fungus / *Aspergillus* / *Bacillus* / *Lactobacillus* / *Pediococcus* / yeast / *Saccharomyces*; **[1]**

(c) reference to amylase; (starch) converted to sugars / eq. / glucose / maltose; **[2]**

(d) (i) anaerobic / low oxygen / temperature of 25–33 °C; **[1]**

(ii) ethanol / alcohol / lactic acid; **[1]**

(Total 6 marks)

Option C Human health and fitness

1 (a) 2.0 μm; **[1]**

(b) A Z line / Z disc / Z band:

B thick filament / myosin;

C A (anisotropic) band / A zone; **[3]**

(c) Z lines shown closer together;

actin filaments, touch / overlap; **[2]**

(Total 6 marks)

2 (a) inflammation of the airways; leads to excessive mucus secretion; obstructs air flow; **[2]**

(b) caused by the bacterium *Mycobacterium tuberculosis*; inhaled in dust / droplets from infected people; colonies of bacteria cause production of tubercles / lesions in the lungs; **[2]**

(c) caused by inhalation of dust particles; particles give rise to fibrosis / increase in the amount of collagen in walls of alveoli; mineral dusts such as silica / asbestos involved; **[2]**

(Total 6 marks)

3 (a) active immunity acquired as a result of exposure to an antigen, while passive immunity is acquired by a child from its mother; child obtains antibodies via placenta / in mother's milk / from colostrum; presence of antigen not needed but antigen is needed in active immunity; antigen triggers production of antibodies and memory cells; **[3]**

(b) T cells can recognise antigens by specific receptors; activated by specific antigens and divide to produce a clone of T cells; some T cells can attach to invading cells and destroy them / activate phagocytosis; attract macrophages; involved in the maturation of B cells; can trigger production of antibodies by B cells; reference to memory T cells ensuring rapid response to subsequent exposure; suppressor cells regulating the immune response; **[3]**

(Total 6 marks)

4 (a) proportional / constant rate of increase from 0 to 400 arbitrary units;

heart stretches more during diastole / contracts more during systole;

peaks / levels off at 500–600 arbitrary units / peaks at 130 cm^3;

decreases at higher work levels / above 600 arbitrary units;

ventricles empty less completely before they refill / during systole; **[3]**

(b) 167–97=72;

72/97 x 100;

=74.23; **[3]**

(c) sharp / steep / eq. rise at first / initially;

then levels off / decreases / rate of increase reduced / peaks at 1000 units / falls after 1000 units;

correct reference to figures;

increases 5 to 5? times between work load 0 and 1200 arbitrary units; **[2]**

(d) controlled by cardiovascular centre / medulla (oblongata);

which receives impulses from baroreceptors / stretch receptors;

in vena cava / right atrium;

impulses from pH receptors / chemoreceptors;

in aorta / aortic arch / carotid sinus;

impulses via sympathetic nervous system;

release of noradrenaline / adrenaline;

reference to effect of high temperature / low pH / low pO_2 / more CO_2;

stretching the cardiac muscle leads to more powerful contraction / eq.;

(Total 12 marks)

Unit 5B Genetics, evolution and biodiversity

1 (a) (grass) not mowed / grazed by domestic animals / livestock / eq.; shrub seedlings allowed to grow / not eaten / eq.; reference to leaf litter from shrubs smothering grasses / *shading* effect; **[2]**

(b) [(19–8)/8] × 100; 137.5 (%); **[2]**

(c) reference to canopy layer in woodland / more layers in woods; increase in number of niches / habitats; increase in different *types* of food; increase in different *types* of nesting / roosting site; **[2]**

(d) more predators (in pine); birds may have larger territories; greater diversity may cause increased competition; decrease in nesting / roosting sites; leaf litter more acidic, supports fewer (insect) species as food for birds / less food available; **[2]**

(e) (secondary) succession; **[1]**

(f) cutting of trees / deforestation / eq.; introduction of new species / aliens / disease (e.g. Dutch elm); natural disaster / flooding / storm damage / volcanic activity / fire / eq.; climate change / acid rain; **[2]**

(Total 11 marks)

2 (a) All / number of, individuals of a species in a, particular area / habitat; At the same time; **[2]**

(b) Number of *Asellus*, more than / double, *Dendrocoelum*, in January / at start of year; both increase from April to July *or* both peak in July; peak number of *Dendrocoelum* more than *Asellus* / figs quoted; both decrease, from July / by August / between July and December; both return to numbers seen / quote figures / at start of year; *Asellus* fluctuates more than *Dendrocoelum* (at any time); **[3]**

(c) General comment about availability of resources for either species e.g. plenty of food / space / less competition for resources / little environmental resistance / inc temperature, leads to inc. reproduction / more vegetation; as *Asellus* population rises *Dendrocoelum* has more food so rises also *or* as *Asellus* falls *Dendrocoelum* has less food so falls also; as *Dendrocoelum* rises *Asellus* is more heavily predated so its numbers fall; *Asellus* runs short of resources when population high, which leads to, decrease / fall, in numbers; **[3]**

(d) specified size of quadrat; (maximum metre squared) refer to type of sampling / how quadrats placed / random sampling; at least 5 quadrats used / running mean; count the number of flatworms found within each quadrat; calculate the mean number of flatworms per m²/eq.; refer to any problems such as difficulty finding flatworms amongst vegetation / only sampling shallow areas / flatworms moving all the time so difficult to count; **[4]**

(Total 12 marks)

3 (a) same time of year / season; reference to migration / breeding / death in winter; same length of observation time; enables all species to be seen; same type / composition of woodland / eq.; same food sources / nesting sites available; same time of day; bird activities vary / ref to nocturnal; same / similar position in wood; species active in different parts of wood; any location / climate reference; valid reason / nesting / food supply; **[4]**

(b) more predatory birds; because more birds to eat / eq.; more nesting / roosting sites; less interspecific competition; more food available; less interspecific competition; more niches / habitats / zones; less competition / more communities; greater variety of plants / trees / eq.; greater variety / range of foods / eq.; **[4]**

(c) more light / less shade; allows growth of other plant species / types / eq.; new / more niches / habitats; greater variety of fruits / seeds / other animals as food for other bird species; different types of nesting sites / eq.; **[3]**

(Total 11 marks)

4 (a) (i) gene is sex linked / Y does not have allele for coat colour / eq.; males inherit X (chromosome) from mother; mother has alleles for orange coat only / mother's (genotype) was OᵒOᵒ / mother homozygous for orange coat / all gametes from mother contain Oᵒ (on X chromosome) / inherit O from mother / eq.; **[3]**

(ii) parental genotypes Oᵇ– and OᵒOᵇ *or* gamete genotypes from male Oᵇ and –, from female Oᵒ and Oᵇ; offspring genotypes Oᵇ– and Oᵒ– and OᵒOᵇ and OᵇOᵇ; black, orange, tortoiseshell; black males, orange males, black females, tortoiseshell females; **[4]**

(b) tyrosinase / enzyme inactive at higher temperatures accept converse;

high / higher, temperature in uterus / womb / mother / or during first few days after birth;

reference to uniform temperature (in uterus) so no markings / same colour (when first born);

after birth / eq. extremities / named extremity are colder than rest of body / eq.;

enzyme activated in, colder regions / extremities / eq.;

reference to change in temperature affecting production of pigment / colour / markings;

correct reference to gene expression;   **[3]**

(c) Manx cat is Mm / heterozygous / Manx cat carries m;

if Manx interbred then chance of tailed offspring / eq.;

credit any diagram showing cross with, Mm and mm / Mm and Mm;

genotype MM, lethal / offspring do not / less likely to survive to birth / die *in utero* / eq.;   **[3]**

(Total 13 marks)

5

Kingdom	Characteristic feature	Representative group
Prokaryotae;	Lack envelope-bound organelles	*Cyanobacteria / Bacteria;*
Protoctista	Possess envelope-bound organelles; often unicells or assemblages of similar cells	*Green algae / brown algae / Protozoa*
Plantae	*Multicellular, photosynthetic with cellulose cell walls;*	Angiosperms
Fungi;	Non-photosynthetic organsims with multinucleate hyphae	Ascomycota, such as *Penicillium*

(Total 5 marks)

6 (a) (distance moved by pigment = 53 or 54 mm

distance moved by solvent = 70 mm)

53 / 70 *or* 5.3 / 7   *or* 54 / 70 *or* 5.4 / 7;

0.76 *or* 0.757   *or* 0.77 *or* 0.771;   **[2]**

(b) chlorophyll a;   **[1]**

(c) photo / cyclic / non-cyclic phosphorylation / light dependent reaction;º

primary pigment / description;

absorbs / traps light (energy) / reference to wavelengths;

emits electrons;

(electrons) passed to, electron / named acceptors / carriers;   **[3]**

(Total 6 marks)

7 (a) P = Ru BP / ribulose biphosphate;

Q = PGA / phosphoglyceric acid / GP / glycerate phosphate;

R = <u>reduced</u> NADP / NADPH / $NADPH_2$ / $NADPH + H^+$;   **[3]**

(b) Calvin cycle;   **[1]**

(c) starch (grains); stroma (of chloroplast);   **[2]**

(Total 6 marks)

8 (a) refer to apical dominance / presence of apical buds inhibits growth of lateral buds / eq.;

apical bud / meristem / tip, is site of auxin synthesis;

auxin, diffuses / moves, down shoot / away from apex;

(auxin) inhibits growth of lateral buds;   **[2]**

(b) both increase in length;

length of buds in B increases faster than those in C / eq.;

begins to slow down after day 5 in B but not in C;

buds in B always longer that those in C;

increase between 0 and 3 > twice as great for B / some comparative figure manipulation;   **[3]**

(c) dormancy of lateral buds broken or reserves the effect of, auxin / antagonistic effect of cytokinin / apical dominance;

(cytokinin) stimulates / promotes, cell division / mitosis (thus growth of buds);   **[2]**

(d) animal hormones produced by, specific organs / glands, but pgs not;

animal hormones produced in, specialised cells / tissues, but pgs not;

animal hormones transported in blood, not so pgs;

(effects of) animal hormones are specific to particular tissues or organs, pgs less specific;

animal hormones have various effects, pgs usually affect growth;   **[3]**

(Total 10 marks)

Unit 5H Genetics, human evolution and biodiversity

1 (a) Aa / X^AX^a / XAXa / eq.; **[1]**

(b) *only* males have the disease / disorder / eq.; (probably) carried on the X chromosomes; male sufferers are produced from unaffected / eq. parents; recessive only expressed when no dominant allele is present / eq. females may be carriers; male has only one locus / allele for the disease / disorder / no locus on Y / eq.; **[3]**

(c) person 3 / father is X^AY / eq.; (as) person 4 / mother is carrier / X^AX^a / Aa / heterozygous; person 5 / son inherits Y from person 3 / father / eq.; and X^a / recessive allele / disorder from person 4 / mother / eq.; **[3]**

(d) person 6 / father is X^AY / A / eq.; person 7 / mother is carrier / eq. is heterozygous / Aa; all the females / daughters will be unaffected / eq.; half / eq. of the sons / males will beunaffected / eq.; there is a *one in four* / eq. chance of producing a sufferer; **[4]**

(e) reference to: amniocentesis / chorionic villus sampling / or description; karyotype made to see chromosome pattern; either of above methods described; DNA extracted and faulty base sequences detected by genetic probe; **[2]**

(Total 13 marks)

2 (a) human bipedal, gorilla quadripedal / partly arboreal / eq.;

(human has) shorter toes / parallel / non-protruding (big) toe, gives flat foot / foot can be placed flat on ground / better balance;

(gorilla has) long toes / separated / opposable (big) toe, for grasping / eq.; **[3]**

(b) bipedal / description;

toes relatively short (compared with foot length);

big toe as long as other toes / toes all similar length;

(big) toe, parallel / close / not opposable, to other toes; **[3]**

(Total 6 marks)

3 knapping; erectus; Neanderthal; cutting / skinning / shearing / sawing; Palaeolithic;

(Total 5 marks)

Index

Page numbers in italics indicate that information is contained in a question and its answer. You will need to look at both pages together.

For example the entry

medulla oblongata *10(107)*

indicates that you need to look at a question on page 10 and its answer on page 107. Note that in this instance the words medulla oblongata only appear on page 107, although information about the medulla oblongata is on both pages.